Handbook of Industrial Hazards and Safety

Handbook of Industrial Hazards and Safety

Paris James

CLANRYE INTERNATIONAL
www.clanryeinternational.com

Clanrye International,
750 Third Avenue, 9th Floor,
New York, NY 10017, USA

ISBN: 978-1-63240-909-6

Cataloging-in-Publication Data

Handbook of industrial hazards and safety / Paris James.
p. cm.
Includes bibliographical references and index.
ISBN 978-1-63240-909-6
1. Industrial safety. 2. Industrial accidents. 3. Safety education, Industrial. I. James, Paris.
T55. H36 2019
660.2804--dc23

For information on all Clanrye International publications visit our website at www.clanryeinternational.com

CLANRYE INTERNATIONAL

Printed in China.

Contents

Preface

Industrial hazards are conditions associated with industrial processes that can potentially cause injury to personnel and can even be life-threatening. These can also cause loss of property or product. Industrial safety is a set of countermeasures against the occupational hazards in an industrial plant such as nuclear plant, oil and gas plant, etc. These countermeasures are aimed at protecting humans, the physical infrastructure and the environment from harm due to the process going beyond the permitted control margins. The use of various safety devices such as flame arresters, safety valves, blowout preventers, chemical protective clothing and insulation monitoring devices can ensure safety in industries. This book is a compilation of chapters that discuss the most vital concepts in the field of industrial hazards and safety. Different approaches, evaluations and methodologies have been included herein. For someone with an interest and eye for detail, this textbook covers the most significant topics in this field.

A detailed account of the significant topics covered in this book is provided below:

Chapter 1- Any condition created by industrial operations that have the capacity to cause potential threat to health, physical injury or damage to property can be termed as an industrial hazard. This chapter has been carefully written to provide an easy understanding of the varied aspects of industrial hazards, the concept of occupational exposure limit, and detailed analysis of occupational hazards and chemical hazards.

Chapter 2- Industrial waste refers to the waste produced due to industrial activity. Some industrial wastes are chemical solvents, industrial by-products, radioactive wastes, etc. This chapter closely analyzes the significant aspects of industrial wastes and the pollution they cause. It includes topics such as chemical waste, hazardous waste, industrial wastewater, agricultural waste, etc.

Chapter 3- The steep rise of industrialization has led to severe impact on the natural and man-made environments. An understanding of the environmental impact of industrial activities requires a study of the major industries of the 21st century namely the plastic industry, construction industry, oil industry, agricultural industry and mining industry. This chapter extensively discusses all such aspects for a comprehensive understanding.

Chapter 4- Any hazard experienced in the workplace is called an occupational hazard. The topics elaborated in this chapter cover the different aspects related to occupation health hazards and safety, such as occupational stress, work accident, occupational injury, occupational noise, occupational heat stress and occupational dust exposure, besides many others.

Chapter 5- Industrial safety devices are designed in order to protect employees from hazardous work conditions. This chapter closely examines the varied industrial safety devices that can be employed, such as blowout preventer, flame arrester, safety valve, chemical protective clothing, insulation monitoring device, etc.

I would like to make a special mention of my publisher who considered me worthy of this opportunity and also supported me throughout the process. I would also like to thank the editing team at the back-end who extended their help whenever required.

Paris James

1

Introduction to Industrial Hazards

Any condition created by industrial operations that have the capacity to cause potential threat to health, physical injury or damage to property can be termed as an industrial hazard. This chapter has been carefully written to provide an easy understanding of the varied aspects of industrial hazards, the concept of occupational exposure limit, and detailed analysis of occupational hazards and chemical hazards.

Hazard is a term associated with a substance that is likely to cause an injury in a given environment or situation. Industrial hazard may be defined as any condition produced by industries that may cause injury or death to personnel or loss of product or property.

Industrial hazards consist of four principle hazards. This is because industries employ many different processes involving a wide range of different raw materials, intermediates, waste products and final products. The hazards encountered are fire, explosion, toxic release and environmental damage.

Fire: This is the most frequent of the hazards however the consequences are generally less. The effect of fire on people usually takes the form of skin burns and is usually dependent on the exposure time and the intensity of the heat. Fire can also produce toxic fumes like Acrolein, Carbon monoxide and Cyanides. Physical structures can be damaged either by the intensity of the heat or combustion. It may also have an effect on essential services like power and instrumentation which can cause an escalation of the incident.

Explosion: Explosions are usually heard from far away as a 'bang'. This is the result of a shock wave. This overpressure can kill people but usually the indirect effects of collapsing buildings, flying glass and debris causes far more loss of life and severe injuries. There are different types of explosions which include gas explosions and dust explosions. Gas explosions occur when a flammable gas mixes with air and is exposed to an ignition source. Dust explosions occur when flammable solids, especially metals, in the form of fine powders are intensively mixed with air and ignited.

Toxic/Chemical release: Sudden releases of toxic vapours have the potential to cause death and severe injuries several miles from the release point. They are carried by water and air. Their release into public sewage systems, rivers, canals and other water courses, either directly or through contaminated water used in firefighting can result in serious threat to public. The number of casualties depends on the weather conditions, population density in the path of the cloud and the effectiveness of the emergency arrangements.

Environmental Damage: As well as having the potential for causing injury, loss of life and damage to property, the hazards of fire, explosion and toxic releases may pose a severe threat to the environment. Release of other substances, not directly toxic to humans can cause major pollution problems. It is becoming increasingly recognized that damage to natural resources such as plant and animal life can have serious long term consequences. E.g. destruction of trees is increasing the effect of global warming and extinction of animals are severely disrupting food webs and causing an increase in pests.

Reducing Risks

- Design and Pre-modification review: this involves proper layout, facilities and material selection. Research should be done try to substitute extremely toxic chemicals with safer ones. Less chemicals should be stored; a reduction in inventory will automatically mean less damage if an accident is to occur.
- Chemical Risk Assessment: Chemicals are assessed based on compatibility, flammability, toxicity, explosion hazards and storage.
- Process Safety Management: HAZOP studies, reliability assessment of process equipment, incorporating safety trips and interlocks, scrubbing system, etc. should be done before effecting major process changes. Management should try to develop a culture of safety in industrial organizations.
- Safety Audits: Periodical assessment of safety procedures and practices, performance of safety systems and gadgets along with follow up measures should be carried out.
- Emergency Planning: A comprehensive risk analysis indicating the impact of consequences and specific written down and practiced emergency procedures along with suitable facilities should be done. This can be done by communities as well as national or regional corporation authorities.
- Training: Proper training of employees and protective services should be done.

- Special times and escorts for dangerous vehicles.
- Public Cooperation on the road: the public should cooperate with the police and any tankers and heavy duty vehicles to avoid accidents and allow for the shortest possible on road time for dangerous vehicles.
- Public awareness: Everyone should be aware of potential disasters and informed of protective and safety measures. MSDS sheets should be readily available to the public. Cautions must be placed to standout on dangerous household and car care products.
- Proper storage of hazardous Materials: All chemicals and hazardous materials should be kept at proper storage temperature and in locked cupboards away from children and animals. Also, if reactive substances are stored, it should be stored is a watertight container.

Examples of some vulnerabilities associated with industrial hazards:

Improper Location of Communities

Communities like California and Couva are located too close to the Point Lisas industrial estate. If there is an explosion or chemical release, there communities will be severely affected potentially with many deaths and structural damage.

Poor Developmental Planning in Point Lisas

Industries in Point Lisas Industrial estate are located in such a way that it is easy for one failure to cause a domino effect e.g. there is a methanol plant situated approximately 150m for a power generation plant. This power plant can produce a very easy source of ignition for any possible leak that may occur from the methanol plant.

Lack of Knowledge

Many persons in the country and primarily persons close to the industrial estate are unaware of the actual dangers they face on a daily basis. Although it has been said may times, persons continue to ignore this because a disaster of catastrophic scale has not occurred before at the estate and companies boast of the low probability of such an incident. From a domestic point of view, person using everyday product like degreasers, disinfectants, bleach, lubricants (WD40, PR40), paints, thinners, acids (concrete cleaner) must be careful as most of these products affect the skin.

Lack Of Mitigation Measures

From findings it has been proven that the mitigation measures put in place on the estate with respect to the ammonia plants are insufficient. There exists no specialized medical facility to deal with industrial cases. Even though there is the Couva Medical facility nearby, this is occupied by persons from that area.

Lack Of Evacuation Expertise

It was found that the emergency response system at Point Lisas was inadequate to handle industrial emergencies as there were a lack of specialized medical personnel as well as triage equipment and facilities.

Transportation Risks

Many chemicals, including flammable hydrocarbons are transported on the roadway alongside other vehicles and pedestrians. This poses a risk of explosion, fire, blast fragments and other harmful injury to bystanders, if an incident was to happen.

Occupational Hazards

Occupational hazard is the risk, harm, or danger that an individual is exposed to at the workplace, whereas occupational diseases result from such exposures to the individual. Although these occupational diseases appear to occur less frequently than other major debilitating diseases, there is evidence that they affect a considerable number of people, particularly in rapidly industrializing countries (e.g., Nigeria), hence indirectly impacting on the economy. During work periods, workers are faced with a variety of hazards almost as numerous as the different types of work, including chemicals, biological agents, physical factors, and adverse ergonomic conditions. These are responsible for a variety of health consequences.

Chemical substances, and their derivatives, are widely used in many sectors including industry, agriculture, mining, water purification, public health—particularly disease eradication—and infrastructure development. Their utilization has brought immense benefits to mankind. However, the production, storage, transportation, and removal of these substances can pose risks to people and the environment, and at the same time it has had negative impacts on human health and safety.

Types of Occupational Hazards

1. Safety

Safety hazards encompass any type of substance, condition or object that can injure workers. In many types of workplaces they can include spills on floors, walkways blocked by cords or boxes, falls from heights, machinery with moving parts, confined spaces and electrical hazards such as frayed cords.

2. Chemical

Workers can be exposed to chemicals in liquids, gases, vapors, fumes and particulate materials. Chemical hazards include acids, pesticides, carbon monoxide, flammable liquids, welding fumes, silica dust and fiberglass fibers.

3. Biological

Employees who work with other people, with animals or with infectious materials can be exposed to biological hazards such as blood, fungi, mold, viruses and animal droppings.

4. Physical

A physical hazard can injure workers with or without contact. These types of hazards include radiation, working in extreme heat or cold, spending hours under the sun or being constantly exposed to loud noise.

5. Ergonomic

Ergonomic related musculoskeletal disorders (MSDs) account for 33% of all employee injury and illness cases. These types of hazards occur when repetitive work, the type of work, or a certain position strains the body. These are the most difficult hazards to spot because problems build up over time.

Examples of Occupational Hazards

Solvents used in the paint industry for example have been shown in numerous studies to be the cause for negative health symptoms that include the central and peripheral nervous system as well as other organ systems. Various studies, as shown in table below, reveal the neurobehavioral effects of organic solvents in paints on paint factory workers/painters. The studies show exposure related negative effects, most commonly on tests of psychomotor function and short-term memory.

Table: Neurobehavioral effects of organic solvents in paints on paint factory workers/ painters

Study	Exposed group (N)	Visual performance/motor deficit	Memory deficit	Behavioral symptoms
Bleeker	Paint factory workers (187)	+	+	–
Maizlish	Spray painters (124)	–	+	–
Fiddler	Painters (101)			+
Cherry	Painters (44)	+	+	+
Baker	Painters (186)	+	+	+
Spurgeon	Painters (90)	+	+	

+, Exposure-related effects observed; –, exposure related effects not seen.

Also, some chemicals (organic and inorganic) used in paint industries contain heavy metals with known risks. Usually, the manufacture of paints involves a wide variety of raw materials that contain heavy metals such as lead, cadmium, and chromium pigments, and fungicides such as mercuric oxide in the production process, which can present with medical hazards, some of which are easily recognized and others that may remain undetected for many years . Lead and mercury, for example, have a serious and irreversible impact on the mental development of children. Exposure to heavy metals has been shown to be associated with middle-term and long-term health risks such as abdominal pain and illness to the human fetus (causing abortion and/or preterm labor). Adults may also experience high blood pressure, fatigue, kidney, and brain disturbances. Chronic exposure to heavy metals may also lead to skin eruptions, intestinal ulcer, and different types of cancers.

To control these medical hazards, particularly in relation to heavy metals, there are coordinated safety initiatives introduced to limit heavy metal exposures in the paint industry thereby preventing negative health effects. These measures include redesigning processes to place a barrier between workers and the hazard; adopting standard operating procedures or safe work practices; providing appropriate training, instruction, or information to reduce the potential for harm and/or adverse health effects to person(s); and implementing the use of personal protective devices (PPDs) such as gloves, glasses, aprons, safety footwear, and dust masks designed to reduce exposure to the hazard. PPDs are usually the last line of defense and usually used in conjunction with one or more of the other control measure.

The consequences of not following these practice guidelines can be fatal as control of these hazards is the key to reducing the risk of injury and illness among workers in this industry.

To effectively control and prevent the occurrence of occupational hazards, employers should:

- Involve workers, who often have the best understanding of the conditions that create hazards and insights into how they can be controlled.
- Identify and evaluate options for controlling hazards, using a "hierarchy of controls."
- Use a hazard control plan to guide the selection and implementation of controls, and implement controls according to the plan.
- Develop plans with measures to protect workers during emergencies and non-routine activities.

Evaluate the effectiveness of existing controls to determine whether they continue to provide protection, or whether different controls may be more effective. Review new technologies for their potential to be more protective, more reliable, or less costly.

Chemical Hazards

Hazardous chemicals in the workplace are substances, mixtures and materials that can be classified according to their health and physicochemical risks and dangers. Physicochemical hazards generally result from a substance's physical and chemical properties, as is the case with flammable, corrosive, oxidizing or explosive substances.

Hazardous chemicals are frequently used in the workplace as raw materials, solvents, cleaning agents, catalysts, and for a number of other functions. These are normally classified according to the risk they pose to health and property.

Hazardous Chemicals Include

Flammable or explosive (e.g. petroleum, TNT, plastic explosives)

Irritating or corrosive to skin, lungs, and eyes (e.g. acids, alkali, paints, fumes)

Toxic chemicals (e.g. carbon monoxide, hydrogen sulfide, cyanide)

Extreme cautions are to be taken when handling, storing, transporting, and using hazardous chemicals. The user needs to wear protective clothing and personal protective equipment to protect themselves from hazardous chemicals. Depending on the chemicals, this might involve rubber gloves, safety goggles, chemical-resistant coveralls, or respirators.

Hazardous chemicals are accompanied by Safety Data Sheets (SDS) that outlines the severity of the potential risks as well as the safety measures that must be in place when handling these substances.

Types of Chemical Hazards

While the approach to defining hazardous chemicals varies by regulatory agency, consider these four basic types of chemical hazards:

Hazard Types	Definition	Examples
Flammable 	Material that will burn or ignite, causing fire or combustion. An ignitable chemical has a flashpoint less than 100° F. A combustible material will burn, but require a flame or elevated temperature plus a spark to start them; and has a flashpoint greater than 100° F but less than 200° F.	Flammables: methanol, acetonitrile, spray adhesive/mount Combustible: diesel fuel, mineral spirits
Corrosive 	Chemical that causes visible destruction of, or irreversible alterations in, living tissue by chemical action at the site of contact. $pH < 2$ and $pH > 12.5$	acetic acid, sodium hydroxide, photographic fixer

Hazard Types	Definition	Examples
Reactive  	Material that reacts violently or explodes under either ambient conditions or when in contact with air, water, or other chemicals. Oxidizers: materials that react strongly with organic materials, sometimes strongly enough to start fires Organic Peroxides: form friction and shock-sensitive explosives Water Reactive: react violently with water Air Reactive (pyrophoric): react violently with air Explosive: designed to explode violently	Oxidizers: nitric acid Organic Peroxides:benzoyl peroxide, methyl ethyl ketone peroxide Water Reactive:sodium metal, sodium borohydride Air Reactive: silane, t-butyl lithium Explosive: TNT, picric acid
Toxic  	Material that may cause harm to an individual if it enters the body. Carcinogen: a substance or agent that may cause cancer Mutagen: An agent that can induce or increase the frequency of mutation in an organism Poison: any substance that can impair function, cause structural damage, or otherwise injure the body Sensitizer: a substance that causes hypersensitivity or reactivity to an antigen, such as pollen, especially by a second or repeated exposure. Teratogen: An agent that causes malformation of an embryo or fetus.	Carcinogen:benzene, carbon tetrachloride Mutagen: bromine Poison: sodium azide, powdered pigments and inks (may contain toxic metals such as chromium and barium) Sensitizer:formaldehyde, phenol Teratogen: PCBs, mercury

Hazard Types	Definition	Examples
Irritant 	Material that can cause harm to an individual in the following ways: Irritant: a substance that can irritate the skin or eyes Skin Sensitizer: a substance which can cause an allergic response following skin contact Acute Toxicity (harmful): a substance that may be fatal or cause organ damage from a single short-term exposure Narcotic Effect: A substance that can cause drowsiness, lack of coordination, and dizziness Hazardous to Ozone Layer (Non-Mandatory)	Powdered substances often have the irritant symbol.
Environmental Hazard 	Toxic to aquatic animals.	Oils and oily debris can be in this class of hazards.

Routes of Entry

Routes of entry describe the way in which a hazardous material enters the body. In a laboratory, the primary route of entry is through inhalation and dermal contact. With proper ventilation such as a chemical fume hoods and proper personal protective equipment, like nitrile gloves, you can reduce the risk of exposure to hazardous chemicals. Good hygiene habits, such as washing your hands after using hazardous chemicals, and using mechanical means to pick up sharps, will reduce your risk of exposure through ingestion or injection. Make sure to report any exposure, or possible exposure, or Risk Management & Safety.

Absorption

Absorption may occur when not using proper engineering controls, work practices/procedures, and/or personal protective equipment (PPE). It may also occur when handling containers, etc. that have been contaminated without wearing PPE.

Reducing your Risk of Exposure through Injection:

- Wear appropriate gloves and other protective gear.
- Remove PPE when entering common spaces and hallways.
- Well label areas that require PPE.
- In the event of an accidental exposure, wash the affected area with soap and water and seek medical attention.

Inhalation

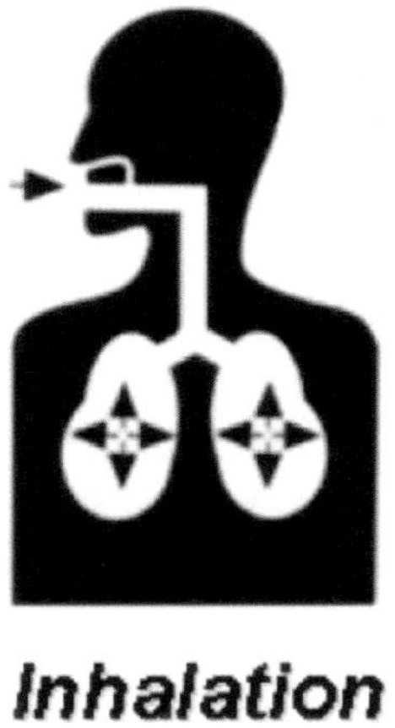

Inhalation may occur when working with volatile materials without appropriate engineering controls, work practices, and/or PPE.

Reducing your Risk of Exposure through Injection:

- Use appropriate engineering controls, work practices, and protective gear to control for the inhalation hazards.
- In the event of an accidental inhalation seek medical attention for instructions.

Ingestion

The gastrointestinal tract is another possible route of entry for hazardous materials. Although direct ingestion of laboratory substances is unlikely, exposure can occur as a result of ingesting contaminated food or beverages, touching the mouth with contaminated fingers or pens, or swallowing inhaled particles.

Reduce your Risk of Exposure through Ingestion:

- Do not eat, drink, smoke, or store food or drink in labs.
- Wash hands thoroughly after working with hazardous materials, even when gloves are worn.
- In the event of accidental ingestion, seek medical care for instructions.
- Do not induce vomiting unless directed to do so by a health care provider.

Injection

Injection effectively bypasses the protection provided by intact skin and provides direct access to the bloodstream, and thus, to internal organ systems. Injection may occur through mishaps with needles, wires, when handling biting animals, or through accidents with broken glass or other sharp objects that have been contaminated.

Reducing your Risk of Exposure through Injection:

- Wear appropriate gloves and other protective gear
- Caution should be taken when working with and around sharp objects
- In the event of an accidental injection, wash the affected area with soap and water and seek medical attention.

Long-Term Risks

Cardiovascular Disease

A 2017 SBU report found evidence that workplace exposure to silica dust, engine exhaust or welding fumes is associated with heart disease. Associations also exist for exposure to arsenic, benzopyrenes, lead, dynamite, carbon disulphide, carbon monoxide, metalworking fluids and occupational exposure to tobacco smoke. Working with the electrolytic production of aluminum, or the production of paper when the sulphate pulping process is used, is associated with heart disease. An association was also found between heart disease and exposure to compounds which are no longer permitted in certain work environments, such as phenoxy acidscontaining TCDD (dioxin) or asbestos.

Workplace exposure to silica dust or asbestos is also associated with pulmonary heart disease. There is evidence that workplace exposure to lead, carbon disulphide, or phenoxyacids containing TCDD, as well as working in an environment where aluminium is being electrolytically produced, are associated with stroke.

First Aid

In case of emergency, it is recommended to understand first aid procedures in order to minimize any damage. Different types of chemicals can cause a variety of damage. Most sources agree that it is best to rinse any contacted skin or eye with water immediately. Currently, there is insufficient evidence of how long the rinsing should be done, as the degree of impacts will vary for substances such as corrosive chemicals. However, the recommended flush time is as follows:

- 5 minutes - non- to mild irritants
- 15 – 20 minutes - moderate to severe irritants and chemicals that cause acute toxicity

- 30 minutes - most corrosives
- 60 minutes - strong alkalis such as sodium, potassium or calcium hydroxide

Transporting the affected person to a health care facility may be important, depending on condition. In the case that the victim needs to be transported before the recommended flush time, then flushing should be done during the transportation process. Some chemical manufacturers may state the specific type of cleansing agent that is recommended.

Occupational Exposure Limit

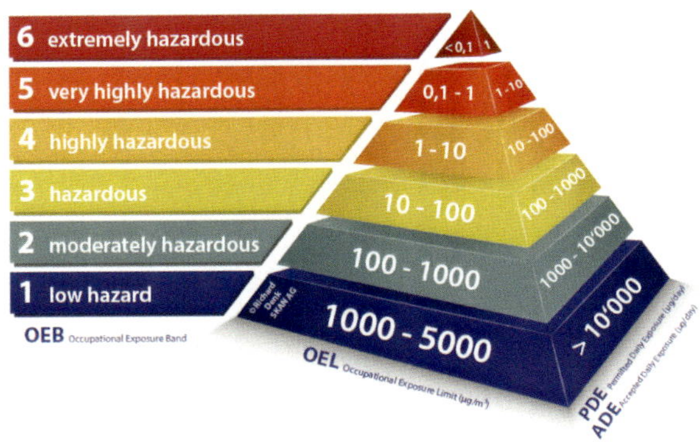

Occupational exposure limits (OELs) are tools to help employers protect the health of those who may be exposed to chemicals in their workplace. Under the United Kingdom Control of Substances Hazardous to Health (COSHH) Regulations, they define adequate control by inhalation. OELs are set by the Health and Safety Commission (HSC) on advice from its Advisory Committee on Toxic Substances (ACTS) and after public consultation. Thus, they are consensus limits which have the support of both sides of industry. COSHH uses two types of occupational exposure limit—the occupational exposure standard (OES) and the maximum exposure limit (MEL).

Types of Exposure Limits

ACGIH defines three categories of threshold limit values:

Threshold Limit Value – Time-Weighted Average (TLV-TWA): The concentration of a hazardous substance in the air averaged over an 8-hour workday and a 40-hour workweek to which it is believed that workers may be repeatedly exposed, day after day, for a working lifetime without adverse effects.

Threshold Limit Value – Short-term exposure (TLV-STEL): A 15-minute time weighted average exposure that should not be exceeded at any time during a workday, even if the overall 8-hour TLV-TWA is below the TLV-TWA. Workers should not be exposed more than four times per day to concentrations between TLV-TWA and TLV-STEL. There should be at least a 60 minute interval between exposures. The short-term exposure threshold has been adopted to account for the acute effects of substances that have primarily chronic affects.

Threshold Limit Value – Ceiling (TLV-C): This is the concentration that should not be exceeded during any part of the working exposure. Peak exposures should be always controlled. For substances that do not have TLV-TWA or TLV-C established, the maximum admissible peak concentrations must not exceed.

Three-times the value of the TLV-TWA for no more than 15 minutes, no more than four times per workday. Exposures must be at least 1 hour apart during the workday.

Five times the TLV-TWA under any circumstances.

The units of measures for the threshold limit values are ppm and mg/m3. The TLVs for aerosols are expressed usually in mg/m3. The TLVs for gases and vapours are expressed in ppm or mg/m3.

In the workplace, a worker may be exposed to several chemical substances at a time. If the toxicological effect of the substances is similar (e.g., each substance affects same target organ or has a similar effect), it can be considered that the combined effect of the chemicals will be the sum of individual effects. A common example is exposure to several organic solvents.

In this case, ACGIH recommends the following calculation:

If the sum of:

$$C1/T1 + C2/T2 + \ldots.Cn/Tn$$

is higher than 1, the threshold limit of the mixture is exceeded.

(C is the concentration in the air of the substance and T is the threshold limit)

This formula should not be used for:

- Mixtures of substances with toxicological effects are not additive (individual toxicological effects and target organs are different).
- Mixtures of substances which inhibit each other's effect.
- Substances that may have a synergistic effect.
- Carcinogens (exposure to mixtures of carcinogens should be eliminated or as low as possible).
- Complex mixtures (e.g., diesel exhaust).

References

- "Occupational health and safety – chemical exposure". www.sbu.se. Swedish Agency for Health Technology Assessment and Assessment of Social Services (SBU). Retrieved 2017-06-01
- Five-workplace-hazards: grainger.com, Retrieved 18 April 2018
- Hazard-prevention, shpguidelines: osha.gov, Retrieved 22 July 2018
- "Hazard symbols and hazard pictograms - Chemical classification". hse.gov.uk. Health and Safety Executive. Retrieved 2016-02-11
- Hazardous-chemical-355: safeopedia.com, Retrieved 14 May 2018

2

Industrial Waste and Pollution

Industrial waste refers to the waste produced due to industrial activity. Some industrial wastes are chemical solvents, industrial by-products, radioactive wastes, etc. This chapter closely analyzes the significant aspects of industrial wastes and the pollution they cause. It includes topics such as chemical waste, hazardous waste, industrial wastewater, agricultural waste, etc.

We have millions of factories, mills, industries, mining plants, etc. around the world. These industries use raw materials to produce finished goods for consumers. But in the manufacturing process, there are materials which are rendered useless. They constitute the industrial waste.

The types of industrial waste generated include cafeteria garbage, dirt and gravel, masonry and concrete, scrap metals, trash, oil, solvents, chemicals, weed grass and trees, wood and scrap lumber, and similar wastes.

Industrial waste has been a problem since the industrial revolution. Industrial waste may be toxic, ignitable, corrosive or reactive. If improperly managed, this waste can pose dangerous health and environmental consequences. In the United States, the amount of hazardous waste generated by manufacturing industries in the country has increased from an estimated 4.5 million tons annually after World War II to some 57 million tons by 1975. By 1990, this total had shot up to approximately 265 million tons. This waste is generated at every stage in the production process, use and disposal of manufactured products. Thus, the introduction of many new products for the home and office - computers, drugs, textiles, paints and dyes, plastics - also introduced hazardous waste, including toxic chemicals, into the environment. These, too, must be managed with extreme care to avoid adverse environmental or human health impacts. The EPA estimated in 1980 that more than 70,000 different chemicals were being manufactured in the U.S., with some 1,000 new chemicals being added each year. The human health and environmental impacts of many of these chemicals are largely unknown. High levels of toxic contaminants have been found in animals and humans, particularly those, like farm workers and oil and gas workers, who are continually exposed to such waste streams. Waste water from manufacturing or chemical processes in industries contributes to water pollution. Industrial waste water usually contains specific and readily identifiable chemical compounds. Water pollution is concentrated within a few subsectors, mainly in the form of toxic wastes and organic pollutants.

Types of Industrial Waste

Industrial waste can be categorized into biodegradable and non-biodegradable.

- Biodegradable: Those industrial wastes which can be decomposed into the non-poisonous matter by the action of certain microorganisms are the biodegradable wastes. They are even comparable to house wastes. These kinds of waste are generated from food processing industries, dairy, textile mills, slaughterhouses, etc. Some examples are paper, leather, wool, animal bones, wheat, etc. They are not toxic in nature and they do not require special treatment either. Their treatment processes include combustion, composting, gasification, bio-methanation, etc.
- Non-biodegradable: Those industrial wastes which cannot be decomposed into non-poisonous substances are the non-biodegradable wastes. Examples are plastics, fly ash, synthetic fibers, gypsum, silver foil, glass objects, radioactive wastes, etc. They are generated by iron and steel plants, fertilizer industries, chemical, drugs, and dyes industries. It is estimated that about 10 to 15 percentage of the total industrial wastes are non-biodegradable and hazardous, and the rate of increase in this category of waste is only increasing every year. These wastes cannot be broken down easily and made less harmful. Hence,

they pollute the environment and cause threat to living organisms. They accumulate in the environment and enter the bodies of animals and plants causing diseases. However, with the advancement in technology, several disposals, and reuse methods have been developed. Wastes from one industry are being treated and utilized in another industry. For example, the cement industry uses the slag and fly ash generated as waste by steel industries. Landfill and incineration are other methods which are being resorted to, for the treated of hazardous wastes.

In Thailand the roles in Municipal solid waste (MSW) management and industrial waste management are organized by the Royal Thai Government, which is then divided into central government, regional government, and local government. Each government is responsible for different tasks. The central government is responsible for stimulating regulation, policies, and standards. The regional governments are responsible for coordinating the central and local governments. The local governments are responsible for waste management in their governed area. However, the local governments do not dispose of the waste by themselves but instead hire private companies that have been granted the right from the Pollution Control Department (PCD) in Thailand. The main companies are Bangpoo Industrial Waste Management Center, General Environmental Conservation Public Company Limited (GENCO), SGS Thailand, Waste Management Siam LTD (WMS), and Better World Green Public Company Limited (BWG). These companies are responsible for the waste they have received from their customers before releasing it to the environment, burying it.

The 1976 Resource Conservation and Recovery Act (RCRA) provides for federal regulation of waste while allowing for individual state control through EPA approved waste disposal programs.

Chemical Waste

Chemical waste is any type of waste that is composed of noxious, potentially hazardous chemicals. Harmful chemicals and solvents that are the byproducts of large scale laboratories and manufacturing plants serve as the most common examples of industrial chemical waste. However, certain household refrigerants, batteries and cleaning products qualify as chemical waste, too. Depending on the potency of certain chemicals and solvents, as well as the potential safety hazards they present, they may fall under the category of hazardous waste.

Disposal

Because of the safety hazards associated with this type of waste, most chemical waste must be disposed of in a special manner. For industrial forms of chemical waste, the disposal process typically involves sealing the waste in securely sealed chemical-resistant drums or barrels, then transporting the safely stored waste to a special landfill. For household forms of chemical waste, many municipal landfills feature areas specifically designated for this type of waste. Like many forms of hazardous waste, chemical waste tends to be toxic, flammable, corrosive or reactive-or some combination thereof-so it's important to take the proper safety precautions when disposing of it.

List of some common chemical wastes

- Unused and surplus reagent grade chemicals
- Intermediates and by-products generated from research & educational experiments
- Batteries
- Anything contaminated by chemicals
- Used oil of all types
- Spent solvents - including water based
- Mercury containing items
- Photographic film processing solutions and chemicals
- Pesticides
- Non-returnable gas cylinders
- Non-empty aerosol cans
- Chemically contaminated sharps
- Finely divided powders
- Contaminated syringes, needles, GC syringes, razor blades, pasteur pipettes, pipette tips

- Equipment and apparatus containing hazardous waste
- Computer/electronic equipment
- Toner cartridges
- Ethylene glycol
- Paints - both oil and latex
- Fluorescent light bulbs
- Light ballasts
- Preserved specimens
- Custodial and industrial cleaners
- Uncured Resins (Phenolic, Epoxy, Styrene, etc.)
- Dye and glazes
- Degreasing solvents
- Brake/Transmission/Power Steering Fluids

Chemical Compatibility Guideline

Many chemicals may react adversely when combined. It is recommended that incompatible chemicals be stored in separate areas of the lab.

Acids should be separated from alkalis, metals, cyanides, sulfides, azides, phosphides, and oxidizers. The reason being, when combined acids with these type of compounds, violent exothermic reaction can occur possibly causing flammable gas, and in some cases explosions.

Oxidizers should be separated from acids, organic materials, metals, reducing agents, and ammonia. This is because when combined oxidizers with these type of compounds, inflammable, and sometimes toxic compounds can occur.

Container Compatibility

When disposing hazardous laboratory chemical waste, chemical compatibility must be considered. For safe disposal, the container must be chemically compatible with the material it will hold. Chemicals must not react with, weaken, or dissolve the container or lid. Acids or bases should not be stored in metal. Hydrofluoric acid should not store in glass. Gasoline (solvents) should not store or transport in lightweight polyethylene containers such as milk jugs. Moreover, the Chemical Compatibility Guidelines should be considered for more detailed information.

Laboratory Waste Containers

Packaging, labeling, storage are the three requirements for disposing chemical waste.

Packaging

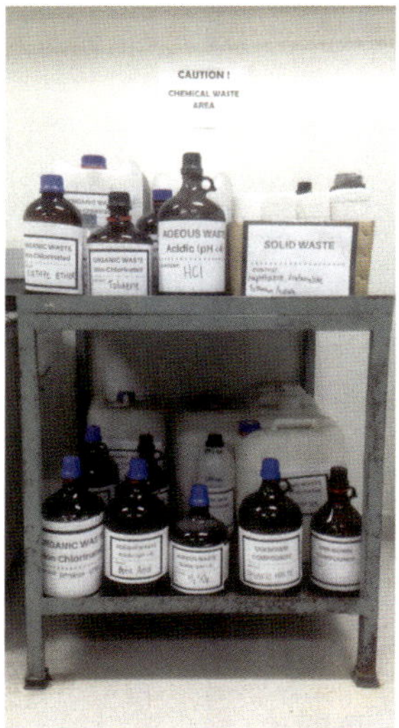

How to properly label, package, and store chemical waste safely.

For packaging, chemical liquid waste containers should only be filled up to 75% capacity to allow for vapour expansion and to reduce potential spills which could occur from moving overfilled containers. Container material must be compatible with the stored hazardous waste. Finally, wastes must not be packaged in containers that improperly identify other non-existing hazards.

In addition to the general packaging requirements mentioned above, incompatible materials should never be mixed together in a single container. Wastes must be stored in containers compatible with the chemicals stored as mentioned in the container compatibility section. Solvent safety cans should to be used to collect and temporarily store large volumes (10–20 liters) of flammable organic waste solvents, precipitates, solids or other non-fluid wastes should not be mixed into safety cans.

Labelling

Label all containers with the group name from the chemical waste category and an itemized list of the contents. All chemicals or anything contaminated with chemicals posing a significant hazard. All waste must be appropriately packaged.

Storage

When storing chemical wastes, the containers must be in good condition and should remain closed unless waste is being added. Hazardous waste must be stored safely prior to removal from the laboratory and should not be allowed to accumulate. Container should be sturdy and leakproof, also has to be labeled. All liquid waste must be stored in leak-proof containers with a screw- top or other secure lid. Snap caps, mis-sized caps, para-film and other loose fitting lids are not acceptable. If necessary, transfer waste material to a container that can be securely closed. Keep waste containers closed except when adding waste. Secondary containment should be in place to capture spills and leaks from the primary container, segregate incompatible hazardous wastes, such as acids and bases.

TOXMAP is a Geographic Information System (GIS) from the Division of Specialized Information Services of the United States National Library of Medicine (NLM) that uses maps of the United States to help users visually explore data from the United States Environmental Protection Agency's (EPA) Toxics Release Inventory and Superfund Basic Research Programs. TOXMAP is a resource funded by the US Federal Government. TOXMAP's chemical and environmental health information is taken from NLM's Toxicology Data Network (TOXNET) and PubMed, and from other authoritative sources.

Chemical waste in our oceans is becoming a major issue for marine life. There have been many studies conducted to try and prove the effects of chemicals in our oceans. In Canada, many of the studies concentrated on the Atlantic provinces, where fishing and aquaculture are an important part of the economy. In New Brunswick, a study was done on the sea urchin in an attempt to identify the effects of toxic and chemical waste on life beneath the ocean, specifically the waste from salmon farms. Sea urchins were used to check the levels of metals in the environment. It is advantageous to use green sea urchins, Strongylocentrotus droebachiensis, because they are widely distributed, abundant in many locations, and easily accessible. By investigating the concentrations of metals in the green sea urchins, the impacts of chemicals from salmon aquaculture activity could be assessed and detected. Samples were taken at 25-metre intervals along a transect in the direction of the main tidal flow. The study found that there was impacts to at least 75 m based on the intestine metal concentrations. So based on this study it is clear that the metals are contaminating the ocean and negatively affecting aquatic life.

Green Sea Urchin or S. droebacheinsis

Another issue regarding chemical waste is the potential risk of surface and groundwater contamination by the heavy metals and radionuclides leached from uranium waste-rock piles (UWRP) A Radionuclide is an atom that has excess nuclear energy, making it unstable. Uranium waste-rock piles refers to Uranium mining, which is the process of extraction of uranium ore from the ground. An example of such threats is in Saskatchewan, Uranium mining and ore processing (milling) can pose a threat to the environment. In open pit mining, large amounts of materials are excavated and disposed off in waste-rock piles. Waste-rock piles from the Uranium mining industry can contain several heavy metals and contaminants that may become mobile under certain conditions. Environmental contaminants may include acid mine drainage, higher concentrations of radionuclides, and non-radioactive metals/metalloids (i.e. As, Mo, Ni, Cu, Zn).

The leachability of heavy metals and radionuclide from UWRP plays a significant role in determining their potential environmental risks to surrounding surface and groundwater. Substantial differences in the solid-phase partitioning and chemical leachability of Ni and U were observed in the investigated UWRP lithological materials and background organic-rich lake sediment. For Instance, in the uranium-mining district of Northern Saskatchewan, Canada, the sequential extraction results showed that a significant amount of Ni (Nickel) was present in the non-labile residual fraction, while Uranium was mostly distributed in the moderately labile fractions. Although Nickel was much less labile than Uranium, the observed Nickel exceeded Uranium concentrations in leaching]]. The observed Nickel and Uranium concentrations were relatively high in the underlying organic-rich lake sediment. Expressed as the percentage of total metal content, potential leachability decreased in the order U > Ni. Data suggest that these elements could potentially migrate to the water table below the UWRP. Detailed information regarding the solid-phase distribution of contaminants in the UWRP is critical to understand the potential for their environmental transport and mobility.

Hazardous Waste

Simply defined, a hazardous waste is a waste with properties that make it dangerous or capable of having a harmful effect on human health or the environment. Hazardous waste is generated from many sources, ranging from industrial manufacturing process wastes to batteries and may come in many forms, including liquids, solids gases, and sludges.

A substance is chemically reactive when it is unstable or could react when exposed to another compound. Hazardous waste that is considered chemically reactive is likely to explode or produce harmful fumes when exposed to other compounds. A flammable substance is anything that is likely to catch on fire, and a corrosive substance is something that can corrode, or break down, metals.

The largest amount of hazardous waste is created in industry, in particular the petroleum and chemical industries, which produce around 70% of all hazardous waste in the United States. The most common examples of hazardous waste from industries in general include solvents used in cleaning, wastewater from petroleum refiners and ash produced from incinerators or coal-burning power plants.

The most dangerous hazardous waste is the waste created by nuclear power plants and nuclear weapons facilities. Although industry creates a large amount of hazardous waste in the United States and other industrial countries, this waste is highly regulated and controlled.

Take a look around your home. Do you think you have any hazardous waste? Although at first you might say no, upon closer inspection most homes contain a variety of hazardous waste items. At any given time, the average household in the United States has approximately 100lbs of hazardous waste on the premises. The most common examples of hazardous waste found within the home include paints, batteries, solvents, cleaning agents and pesticides. Now think about your home again. Do you have any of these items in it?

Each year in the United States, around 1.6 million tons of hazardous waste is disposed of by households. Unfortunately, hazardous waste produced from households is far less regulated than industry and more likely to be disposed of improperly, like down the drain, and have a negative impact on the environment.

Disposal

Being that hazardous waste can be harmful in a variety of ways, it needs to be disposed of differently than non-hazardous waste. There are three main methods for disposing of hazardous waste.

The first method is to put solid hazardous waste in sanitary landfills, which are a method of waste disposal where the waste is buried either underground or in large piles. Although non-hazardous waste is often disposed of in landfills, the landfills for hazardous waste are constructed and monitored differently. Landfills for hazardous waste are made with thicker, impervious liners and with more heavy-duty removal systems for leaching materials. They are also constructed far from aquifers to reduce the risk of water contamination.

Hazardous waste that is liquid or has been dissolved is often placed in surface impoundments, which are shallow depressions in the earth that are lined with plastic and impervious materials. The liquid hazardous waste is dumped in the impoundment and left to evaporate. Once the liquid has evaporated, the solid hazardous waste residue remains at the bottom of the impoundment and can be removed and transported to a landfill. Surface impoundments pose many risks, including contamination, and are only used for temporary processing and storage.

The third method of hazardous waste disposal is deep-well injection, which is when liquid waste is injected into a well that has been created in the porous rock deep below the water table. Around nine billion gallons of hazardous waste are injected into deep-wells each year in the United States. Although this method of hazardous waste disposal is designed to be long-term and keep the waste away from humans and ground water, sometimes the wells leak or are damaged and waste contaminates the water supply.

Hazardous Waste Mapping Systems

The U.S. government provides several tools for mapping hazardous wastes to particular locations. These tools also allow the user to view additional information.

- TOXMAP is a Geographic Information System (GIS) service from the Division of Specialized Information Services of the United States National Library of Medicine(NLM) that uses maps of the United States to help users visually explore data from the United States Environmental Protection Agency's (EPA) Toxics Release Inventory and Superfund Basic Research Program. This is a resource funded by the US Federal Government. TOXMAP's chemical and environmental health information is taken from NLM's Toxicology Data Network (TOXNET), PubMed, and other authoritative sources.
- The US Environmental Protection Agency (EPA) "Where You Live" allows users to select a region from a map to find information about Superfund sites in that region.

Universal Wastes

Universal wastes are a special category of hazardous wastes that (in the U.S.):

- Generally pose a lower threat relative to other hazardous wastes are ubiquitous and produced in very large quantities by a large number of generators.

Some of the most common "universal wastes" are: fluorescent light bulbs, some specialty batteries (e.g. lithium or lead containing batteries), cathode ray tubes, and mercury-containing devices.

Universal wastes are subject to somewhat less stringent regulatory requirements. Small quantity generators of universal wastes may be classified as "conditionally exempt small quantity generators" (CESQGs) which release them from some of the regulatory requirements for the handling and storage hazardous wastes.

Universal wastes must still be disposed of properly.

Household Hazardous Waste

Household Hazardous Waste (HHW), also referred to as domestic hazardous waste or home generated special materials, is a waste that is generated from residential households. HHW only applies to waste coming from the use of materials that are labeled for and sold for "home use". Waste generated by a company or at an industrial setting is not HHW.

The following list includes categories often applied to HHW. It is important to note that many of these categories overlap and that many household wastes can fall into multiple categories:

- Paints and solvents
- Automotive wastes (used motor oil, antifreeze, etc.)
- Pesticides (insecticides, herbicides, fungicides, etc.)
- Mercury-containing wastes (thermometers, switches, fluorescent lighting, etc.)
- Electronics (computers, televisions, cell phones)
- Aerosols / Propane cylinders
- Caustics / Cleaning agents
- Refrigerant-containing appliances
- Some specialty batteries (e.g. lithium, nickel cadmium, or button cell batteries)
- Ammunition
- Radioactive wastes (some home smoke detectors are classified as radioactive waste because they contain very small amounts of radioactive isotope americium.

Disposal of Hazardous Waste

Historically, some hazardous wastes were disposed of in regular landfills. This resulted in unfavorable amounts of hazardous materials seeping into the ground. These chemicals eventually entered to natural hydrologic systems. Many landfills now require countermeasures against groundwater contamination. For example, a barrier has to be installed along the foundation of the landfill to contain the hazardous substances that may remain in the disposed waste. Currently, hazardous wastes must often be stabilized and solidified in order to enter a landfill and must undergo different treatments in order to stabilize and dispose of them. Most flammable materials can be recycled into industrial fuel. Some materials with hazardous constituents can be recycled, such as lead acid batteries.

Recycling

Some hazardous wastes can be recycled into new products. Examples may include lead-acid batteries or electronic circuit boards. When heavy metals in these types of ashes go through the proper treatment, they could bind to other pollutants and convert them into easier-to-dispose solids, or they could be used as pavement filling. Such treatments reduce the level of threat of harmful chemicals, like fly and bottom ash, while also recycling the safe product.

Portland Cement

Another commonly used treatment is cement based solidification and stabilization. Cement is used because it can treat a range of hazardous wastes by improving physical characteristics and decreasing the toxicity and transmission of contaminants. The cement produced is categorized into 5 different divisions, depending on its strength and components. This process of converting sludge into cement might include the addition of pH adjustment agents, phosphates, or sulfur reagents to reduce the settling or curing time, increase the compressive strength, or reduce the leach ability of contaminants.

Incineration, Destruction and Waste-To-Energy

Hazardous waste may be "destroyed". For example, by incinerating it at a high temperature, flammable wastes can sometimes be burned as energy sources. For example, many cement kilns burn hazardous wastes like used oils or solvents. Today, incineration treatments not only reduce the amount of hazardous waste, but also generate energy from the gases released in the process. It is known that this particular waste treatment releases toxic gases produced by the combustion of byproduct or other materials which can affect the environment. However, current technology has developed more efficient incinerator units that control these emissions to a point where this treatment is considered a more beneficial option. There are different types of incinerators which vary depending on the characteristics of the waste. Starved air incineration is another method used to treat hazardous wastes. Just like in common incineration, burning occurs, however controlling the amount of oxygen allowed proves to be significant to reduce the amount of harmful byproducts produced. Starved air incineration is an improvement of the traditional incinerators in terms of air pollution. Using this technology, it is possible to control the combustion rate of the waste and therefore reduce the air pollutants produced in the process.

Hazardous Waste Landfill

Hazardous waste may be sequestered in an hazardous waste landfill or permanent disposal facility. "In terms of hazardous waste, a landfill is defined as a disposal facility or part of a facility where hazardous waste is placed or on land and which is not a pile, a land treatment facility, a surface impoundment, an underground injection well, a salt dome formation, a salt bed formation, an underground mine, a cave, or a corrective action management unit."

Pyrolysis

Some hazardous waste types may be eliminated using pyrolysis in an ultra high temperature electrical arc, in inert conditions to avoid combustion. This treatment method may be preferable to high temperature incineration in some circumstances such as in the destruction of concentrated organic waste types, including PCBs, pesticides and other persistent organic pollutants.

Hazardous-waste management is the collection, treatment, and disposal of waste material that, when improperly handled, can cause substantial harm to human health and safety or to the environment. Hazardous wastes can take the form of solids, liquids, sludges, or contained gases, and they are generated primarily by chemical production, manufacturing, and other industrial activities. They may cause damage during inadequate storage, transportation, treatment, or disposal operations. Improper hazardous-waste storage or disposal frequently contaminates surface and groundwater supplies. People living in homes built near old and abandoned waste disposal sites may be in a particularly vulnerable position. In an effort to remedy existing problems and to prevent future harm from hazardous wastes, governments closely regulate the practice of hazardous-waste management.

Hazardous wastes are classified on the basis of their biological, chemical, and physical properties. These properties generate materials that are either toxic, reactive, ignitable, corrosive, infectious, or radioactive.

Toxic wastes are poisons, even in very small or trace amounts. They may have acute effects, causing death or violent illness, or they may have chronic effects, slowly causing irreparable harm. Some are carcinogenic, causing cancer after many years of exposure. Others are mutagenic, causing major biological changes in the offspring of exposed humans and wildlife.

Reactive wastes are chemically unstable and react violently with air or water. They cause explosions or form toxic vapours. Ignitable wastes burn at relatively low temperatures and may cause an immediate fire hazard. Corrosive wastes include strong acidic or alkaline substances. They destroy solid material and living tissue upon contact, by chemical reaction.

Infectious wastes include used bandages, hypodermic needles, and other materials from hospitals or biological research facilities. Radioactive wastes emit ionizing energy that can harm living organisms. Because some radioactive materials can persist in the environment for many thousands of years before fully decaying, there is much concern over the control of these wastes. However, the handling and disposal of radioactive material is not a responsibility of local municipal government. Because of the scope and complexity of the problem, the management of radioactive waste—particularly nuclear fission waste—is usually considered an engineering task separate from other forms of hazardous-waste management.

Transport of Hazardous Waste

Hazardous waste generated at a particular site often requires transport to an approved treatment, storage, or disposal facility (TSDF). Because of potential threats to public safety and the environment, transport is given special attention by governmental agencies. In addition to the occasional accidental spill, hazardous waste has, in the past, been intentionally spilled or abandoned at random locations in a practice known as "midnight dumping." This practice has been greatly curtailed by the enactment of laws that require proper labeling, transport, and tracking of all hazardous wastes.

Transport Vehicles

Hazardous waste is generally transported by truck over public highways. Only a very small amount is transported by rail, and almost none is moved by air or inland waterway. Highway shipment is the most common because road vehicles can gain access to most industrial sites and approved TSDFs. Railroad trains require expensive siding facilities and are suitable only for very large waste shipments.

Hazardous wastes can be shipped in tank trucks made of steel or aluminum alloy, with capacities up to about 34,000 litres (9,000 gallons). They also can be containerized and shipped in 200-litre (55-gallon) drums. Specifications and standards for cargo tank trucks and shipping containers are included in governmental regulations.

The Manifest System

In the United States a key feature of regulations pertaining to waste transport is the "cradle-to-grave" manifest system, which monitors the journey of hazardous waste from its point of origin to the point of final disposal. The manifest system helps to eliminate the problem of midnight dumping. It also provides a means for determining the type and quantity of hazardous waste being generated, as well as the recommended emergency procedures in case of an accidental spill. A manifest is a record-keeping document that must be prepared by the generator of the hazardous waste, such as a chemical manufacturer. The generator has primary responsibility for the ultimate disposal of the waste and must give the manifest, along with the waste itself, to a licensed waste transporter. A copy of the manifest must be delivered by the transporter to the recipient of the waste at an authorized TSDF. Each time the waste changes hands, a copy of the manifest must be signed. Copies of the manifest are kept by each party involved, and additional copies are sent to appropriate environmental agencies.

In the event of a leak or accidental spill of hazardous waste during its transport, the transporter must take immediate and appropriate actions, including notifying local authorities of the discharge. An area may have to be diked to contain the wastes, and efforts must be undertaken to remove the wastes and reduce environmental or public health hazards.

Treatment, Storage, and Disposal

Several options are available for hazardous-waste management. The most desirable is to reduce the quantity of waste at its source or to recycle the materials for some other productive use. Nevertheless, while reduction and recycling are desirable options, they are not regarded as the final remedy to the problem of hazardous-waste disposal. There will always be a need for treatment and for storage or disposal of some amount of hazardous waste.

Treatment

Hazardous waste can be treated by chemical, thermal, biological, and physical methods. Chemical methods include ion exchange, precipitation, oxidation and reduction, and neutralization. Among thermal methods is high-temperature incineration, which not only can detoxify certain organic wastes but also can destroy them. Special types of thermal equipment are used for burning waste in either solid, liquid, or sludge form. These include the fluidized-bed incinerator, multiple-hearth furnace, rotary kiln, and liquid-injection incinerator. One problem posed by hazardous-waste incineration is the potential for air pollution.

Biological treatment of certain organic wastes, such as those from the petroleum industry, is also an option. One method used to treat hazardous waste biologically is called land farming. In this technique the waste is carefully mixed with surface soil on a suitable tract of land. Microbes that can metabolize the waste may be added, along with nutrients. In some cases a genetically engineered species of bacteria is used. Food or forage crops are not grown on the same site. Microbes can also be used for stabilizing hazardous wastes on previously contaminated sites; in that case the process is called bioremediation.

The chemical, thermal, and biological treatment methods outlined above change the molecular form of the waste material. Physical treatment, on the other hand, concentrates, solidifies, or reduces the volume of the waste. Physical processes include evaporation, sedimentation, flotation, and filtration. Yet another process is solidification, which is achieved by encapsulating the waste in concrete, asphalt, or plastic. Encapsulation produces a solid mass of material that is resistant to leaching. Waste can also be mixed with lime, fly ash, and water to form a solid, cement like product.

Surface Storage and Land Disposal

Hazardous wastes that are not destroyed by incineration or other chemical processes need to be disposed of properly. For most such wastes, land disposal is the ultimate destination, although it is not an attractive practice, because of the inherent environmental risks involved. Two basic methods of land disposal include landfilling and underground injection. Prior to land disposal, surface storage or containment systems are often employed as a temporary method.

Temporary on-site waste storage facilities include open waste piles and ponds or lagoons. New waste piles must be carefully constructed over an impervious base and must comply with regulatory requirements similar to those for landfills. The piles must be protected from wind dispersion or erosion. If leachate is generated, monitoring and control systems must be provided. Only no containerized solid, no flowing waste material can be stored in a new waste pile, and the material must be landfilled when the size of the pile becomes unmanageable.

A common type of temporary storage impoundment for hazardous liquid waste is an open pit or holding pond, called a lagoon. New lagoons must be lined with impervious clay soils and flexible membrane liners in order to protect groundwater. Leachate collection systems must be installed between the liners, and groundwater monitoring wells are required. Except for some sedimentation, evaporation of volatile organics, and possibly some surface aeration, open lagoons provide no treatment of the waste. Accumulated sludge must be removed periodically and subjected to further handling as a hazardous waste.

Many older, unlined waste piles and lagoons are located above aquifers used for public water supply, thus posing significant risks to public health and environmental quality. A large number of these old sites have been identified and scheduled for cleanup, or remediation.

Secure Landfills

Landfilling of hazardous solid or containerized waste is regulated more stringently than landfilling of municipal solid waste. Hazardous wastes must be deposited in so-called secure landfills, which provide at least 3 meters (10 feet) of separation between the bottom of the landfill and the underlying bedrock or groundwater table. A secure hazardous-waste landfill must have two impermeable liners and leachate collection systems. The double leachate collection system consists of a network of perforated pipes placed above each liner. The upper system prevents the accumulation of leachate trapped in the fill, and the lower serves as a backup. Collected leachate is pumped to a treatment plant. In order to reduce the amount of leachate in the fill and minimize the potential for environmental damage, an impermeable cap or cover is placed over a finished landfill.

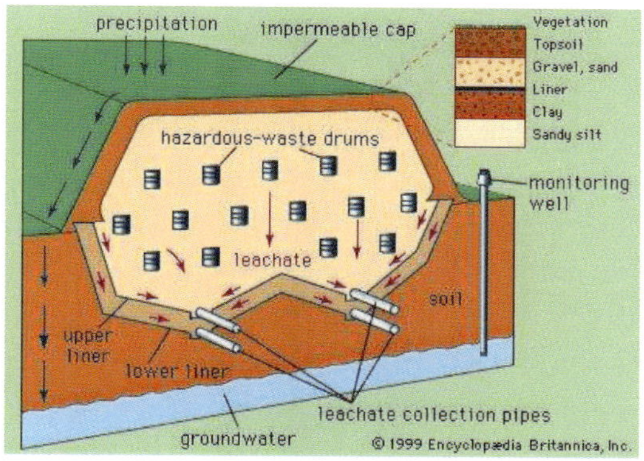

Schematic diagram of a secure hazardous-waste landfill with a double leachate collection system

A groundwater monitoring system that includes a series of deep wells drilled in and around the site is also required. The wells allow a routine program of sampling and testing to detect any leaks or groundwater contamination. If a leak does occur, the wells can be pumped to intercept the polluted water and bring it to the surface for treatment.

One option for the disposal of liquid hazardous waste is deep-well injection, a procedure that involves pumping liquid waste through a steel casing into a porous layer of limestone or sandstone. High pressures are applied to force the liquid into the pores and fissures of the rock, where it is to be permanently stored. The injection zone must lie below a layer of impervious rock or clay, and it may extend more than 0.8 km (0.5 mile) below the surface. Deep-well injection is relatively inexpensive and requires little or no pretreatment of the waste, but it poses a danger of leaking hazardous waste and eventually polluting subsurface water supplies.

Remedial Action

Disposal of hazardous waste in unlined pits, ponds, or lagoons poses a threat to human health and environmental quality. Many such uncontrolled disposal sites were used in the past and have been abandoned. Depending on a determination of the level of risk, it may be necessary to remediate those sites. In some cases, the risk may require emergency action. In other instances, engineering studies may be required to assess the situation thoroughly before remedial action is undertaken.

One option for remediation is to completely remove all the waste material from the site and transport it to another location for treatment and proper disposal. This so-called off-site solution is usually the most expensive option. An alternative is on-site remediation, which reduces the production of leachate and lessens the chance of groundwater contamination. On-site remediation may include temporary removal of the hazardous waste, construction of a secure landfill on the same site, and proper replacement of the waste. It may also include treatment of any contaminated soil or groundwater. Treated soil may be replaced on-site and treated groundwater returned to the aquifer by deep-well injection.

A less costly alternative is full containment of the waste. This is done by placing an impermeable cover over the hazardous-waste site and by blocking the lateral flow of groundwater with subsurface cutoff walls. It is possible to use cutoff walls for this purpose when there is a natural layer of impervious soil or rock below the site. The walls are constructed around the perimeter of the site, deep enough to penetrate to the impervious layer. They can be excavated as trenches around the site without moving or disturbing the waste material. The trenches are filled with a bentonite clay slurry to prevent their collapse during construction, and they are backfilled with a mixture of soil and cement that solidifies to form an impermeable barrier. Cutoff walls thus serve as vertical barriers to the flow of water, and the impervious layer serves as a barrier at the bottom.

Industrial Wastewater

Industrial wastewater is a by-product of industrial or commercial activities. Whether it's the food we eat, the beverages we drink, the clothes we wear, or the paper and chemical products we use, water is required for nearly every step of production across a multitude of different industries. The resulting wastewater must be carefully managed.

Regardless of how wastewater is treated, the "end product" is called effluent. To comply with environmental protection laws, certain things must be removed from the wastewater. This includes organic matter, inorganics (sodium, potassium, calcium, magnesium, copper, lead, nickel, and zinc), pathogens, and nutrients (most notably nitrogen and phosphorus). The treated wastewater can then be safely discharged into water bodies, applied to land, or even reused in plant operations.

Sources Of Industrial Wastewater

Battery Manufacturing

Battery manufacturers specialize in fabricating small devices for electronics and portable equipment (e.g., power tools), or larger, high-powered units for cars, trucks and other motorized vehicles. Pollutants generated at manufacturing plants includes cadmium, chromium, cobalt, copper, cyanide, iron, lead, manganese, mercury, nickel, oil & grease, silver and zinc.

Organic Chemicals Manufacturing

The specific pollutants discharged by organic chemical manufacturers vary widely from plant to plant, depending on the types of products manufactured, such as bulk organic chemicals, resins, pesticides, plastics, or synthetic fibers. Some of the organic compounds that may be discharged are benzene, chloroform, naphthalene, phenols, toluene and vinyl chloride. Biochemical oxygen demand (BOD), which is a gross measurement of a range of organic pollutants, may be used to gauge the effectiveness of a biological wastewater treatment system, and is used as a regulatory parameter in some discharge permits. Metal pollutant discharges may include chromium, copper, lead, nickel and zinc.

Electric Power Plants

Fossil-fuel power stations, particularly coal-fired plants, are a major source of industrial wastewater. Many of these plants discharge wastewater with significant levels of metals such as lead, mercury, cadmium and chromium, as well as arsenic, selenium and nitrogen compounds (nitrates and nitrites). Wastewater streams include flue-gas desulfurization, fly ash, bottom ash and flue gas mercury control. Plants with air pollution controls such as wet scrubbers typically transfer the captured pollutants to the wastewater stream.

Ash ponds, a type of surface impoundment, are a widely used treatment technology at coal-fired plants. These ponds use gravity to settle out large particulates (measured as total suspended solids) from power plant wastewater. This technology does not treat dissolved pollutants. Power stations use additional technologies to control pollutants, depending on the particular waste stream in the plant. These include dry ash handling, closed-loop ash recycling, chemical precipitation, biological treatment (such as an activated sludge process), membrane systems, and evaporation-crystallization systems. Technological advancements in ion exchange membranes and electro dialysis systems has enabled high efficiency treatment of flue-gas desulfurization wastewater to meet recent EPA discharge limits. The treatment approach is similar for other highly scaling industrial wastewaters.

Food Industry

Wastewater generated from agricultural and food operations has distinctive characteristics that set it apart from common municipal wastewater managed by public or private sewage treatment plants throughout the world: it is biodegradable and non-toxic, but has high concentrations of biochemical oxygen demand (BOD) and suspended solids (SS). The constituents of food and agriculture wastewater are often complex to predict, due to the differences in BOD and pH in effluents from vegetable, fruit, and meat products and due to the seasonal nature of food processing and post-harvesting.

Processing of food from raw materials requires large volumes of high grade water. Vegetable washing generates waters with high loads of particulate matter and some dissolved organic matter. It may also contain surfactants.

Animal slaughter and processing produces very strong organic waste from body fluids, such as blood, and gut contents. This wastewater is frequently contaminated by significant levels of antibiotics and growth hormones from the animals and by a variety of pesticides used to control external parasites.

Processing food for sale produces wastes generated from cooking which are often rich in plant organic material and may also contain salt, flavorings, coloring material and acids or alkali. Very significant quantities of oil or fats may also be present.

Food processing activities such as plant cleaning, material conveying, bottling, and product washing create wastewater. Many food processing facilities require on-site treatment before operational wastewater can be land applied or discharged to a waterway or a sewer system. High suspended solids levels of organic particles increase Biochemical Oxygen Demand (BOD) can result in significant sewer surcharge fees. Sedimentation, wedge wire screening, or rotating belt filtration (micro screening) are commonly used methods to reduce suspended organic solids loading prior to discharge.

Iron and Steel Industry

The production of iron from its ores involves powerful reduction reactions in blast furnaces. Cooling waters are inevitably contaminated with products especially ammonia and cyanide. Production of coke from coal in coking plants also requires water cooling and the use of water in by-products separation. Contamination of waste streams includes gasification products such as benzene, naphthalene, anthracene, cyanide, ammonia, phenols, cresols together with a range of more complex organic compounds known collectively as polycyclic aromatic hydrocarbons (PAH).

The conversion of iron or steel into sheet, wire or rods requires hot and cold mechanical transformation stages frequently employing water as a lubricant and coolant. Contaminants include hydraulic oils, tallow and particulate solids. Final treatment of iron and steel products before onward sale into manufacturing includes pickling in strong mineral acid to remove rust and prepare the surface for tin or chromium plating or for other surface treatments such as galvanization or painting. The two acids commonly used are hydrochloric acid and sulfuric acid. Wastewaters include acidic rinse waters together with waste acid. Although many plants operate acid recovery plants (particularly those using hydrochloric acid), where the mineral acid is boiled away from the iron salts, there remains a large volume of highly acid ferrous sulfate or ferrous chloride to be disposed of. Many steel industry wastewaters are contaminated by hydraulic oil, also known as soluble oil.

Mines and Quarries

Mine wastewater effluent in Peru, with neutralized pH from tailing runoff

The principal waste-waters associated with mines and quarries are slurries of rock particles in water. These arise from rainfall washing exposed surfaces and haul roads and also from rock washing and grading processes. Volumes of water can be very high, especially rainfall related arisings on large sites. Some specialized separation operations, such as coal washing to separate coal from native rock using density gradients, can produce wastewater contaminated by fine particulate haematite and surfactants. Oils and hydraulic oils are also common contaminants.

Wastewater from metal mines and ore recovery plants are inevitably contaminated by the minerals present in the native rock formations. Following crushing and extraction of the desirable materials, undesirable materials may enter the wastewater stream. For metal mines, this can include unwanted metals such as zinc and other materials such as arsenic. Extraction of high value metals such as gold and silver may generate slimes containing very fine particles in where physical removal of contaminants becomes particularly difficult.

Additionally, the geologic formations that harbor economically valuable metals such as copper and gold very often consist of sulphide type ores. The processing entails grinding the rock into fine particles and then extracting the desired metal(s), with the leftover rock being known as tailings. These tailings contain a combination of not only undesirable leftover metals, but also sulphide components which eventually form sulphuric acid upon the exposure to air and water that inevitably occurs when the tailings are disposed of in large impoundments. The resulting acid mine drainage, which is often rich in heavy metals (because acids dissolve metals), is one of the many environmental impacts of mining.

Nuclear Industry

The waste production from the nuclear and radio-chemicals industry is dealt with as Radioactive waste.

Petroleum Refining and Petrochemicals

Pollutants discharged at petroleum refineries and petrochemical plants include conventional pollutants (biochemical oxygen demand, oil and grease, suspended solids), ammonia, chromium, phenols and sulfides.

Pulp and Paper Industry

Effluent from the pulp and paper industry is generally high in suspended solids and BOD. Plants that bleach wood pulp for paper making may generate chloroform, dioxins (including 2,3,7,8-TCDD), furans, phenols and chemical oxygen demand (COD). Stand-alone paper mills using imported pulp may only require simple primary treatment, such as sedimentation or dissolved air flotation. Increased BOD or COD loadings, as well as organic pollutants, may require biological treatment such as activated sludge or upflow anaerobic sludge blanket reactors. For mills with high inorganic loadings like salt, tertiary treatments may be required, either general membrane treatments like ultrafiltration or reverse osmosis or treatments to remove specific contaminants, such as nutrients.

Textile Dyeing

Textile dyeing plants generate wastewater that contain synthetic and natural dyestuff, gum thickener (guar) and various wetting agents, pH buffers and dye retardants or accelerators. Following treatment with polymer-based flocculants and settling agents, typical monitoring parameters include BOD, COD, color (ADMI), sulfide, oil and grease, phenol, TSS and heavy metals (chromium, zinc, lead, copper).

Industrial Oil Contamination

Industrial applications where oil enters the wastewater stream may include vehicle wash bays, workshops, fuel storage depots, transport hubs and power generation. Often the wastewater is discharged into local sewer or trade waste systems and must meet local environmental specifications. Typical contaminants can include solvents, detergents, grit. lubricants and hydrocarbons.

Water Treatment

Many industries have a need to treat water to obtain very high quality water for demanding purposes such pure chemical synthesis or boiler feed water. Many water treatment produce organic and mineral sludges from filtration and sedimentation. Ion exchange using natural or synthetic resins removes calcium, magnesium and carbonate ions from water, typically replacing them with sodium, chloride, hydroxyl and/or other ions. Regeneration of ion exchange columns with strong acids and alkalis produces a wastewater rich in hardness ions which are readily precipitated out, especially when in admixture with other wastewater constituents.

Wood Preserving

Wood preserving plants generate conventional and toxic pollutants, including arsenic, COD, copper, chromium, abnormally high or low pH, phenols, oil & grease, and suspended solids.

Wool Processing

Insecticide residues in fleeces are a particular problem in treating waters generated in wool processing. Animal fats may be present in the wastewater, which if not contaminated, can be recovered for the production of tallow or further rendering.

Treatment Of Industrial Wastewater

The various types of contamination of wastewater require a variety of strategies to remove the contamination.

Brine Treatment

Brine treatment involves removing dissolved salt ions from the waste stream. Although similarities to seawater or brackish water desalination exist, industrial brine treatment may contain unique combinations of dissolved ions, such as hardness ions or other metals, necessitating specific processes and equipment.

Brine treatment systems are typically optimized to either reduce the volume of the final discharge for more economic disposal (as disposal costs are often based on volume) or maximize the recovery of fresh water or salts. Brine treatment systems may also be optimized to reduce electricity consumption, chemical usage, or physical footprint.

Brine treatment is commonly encountered when treating cooling tower blow down, produced water from steam assisted gravity drainage (SAGD), produced water from natural gas extraction such as coal seam gas, frac flow back water, acid mine or acid rock drainage, reverse osmosis reject, chlor-alkali wastewater, pulp and paper mill effluent, and waste streams from food and beverage processing.

Brine treatment technologies may include: membrane filtration processes, such as reverse osmosis; ion exchange processes such as electro dialysis or weak acid cation exchange; or evaporation processes, such as brine concentrators and crystallizers employing mechanical vapour recompression and steam.

Reverse osmosis may not be viable for brine treatment, due to the potential for fouling caused by hardness salts or organic contaminants, or damage to the reverse osmosis membranes from hydrocarbons.

Evaporation processes are the most widespread for brine treatment as they enable the highest degree of concentration, as high as solid salt. They also produce the highest purity effluent, even distillate-quality. Evaporation processes are also more tolerant of organics, hydrocarbons, or hardness salts. However, energy consumption is high and corrosion may be an issue as the prime mover is concentrated salt water. As a result, evaporation systems typically employ titanium or steel materials.

Brine Management

Brine management examines the broader context of brine treatment and may include consideration of government policy and regulations, corporate sustainability, environmental impact, recycling, handling and transport, containment, centralized compared to on-site treatment, avoidance and reduction, technologies, and economics. Brine management shares some issues with leachate management and more general waste management.

Solids Removal

Most solids can be removed using simple sedimentation techniques with the solids recovered as slurry or sludge. Very fine solids and solids with densities close to the density of water pose special problems. In such case filtration or ultrafiltration may be required. Although, flocculation may be used, using alum salts or the addition of polyelectrolytes.

Wastewater from industrial food processing often requires on-site treatment before it can be discharged to prevent or reduce sewer surcharge fees. The type of industry and specific operational practices determine what types of wastewater is generated and what type of treatment is required. Reducing solids such as waste product, organic materials, and sand is often a goal of industrial wastewater treatment. Some common ways to reduce solids include primary sedimentation (clarification), Dissolved Air Flotation or (DAF), belt filtration (micro screening), and drum screening.

Oils and Grease Removal

The effective removal of oils and grease is dependent on the characteristics of the oil in terms of its suspension state and droplet size, which will in turn affect the choice of separator technology. Oil in industrial waste water may be free light oil, heavy oil, which tends to sink, and emulsified oil, often referred to as soluble oil. Emulsified or soluble oils will typically required "cracking" to free the oil from its emulsion. In most cases this is achieved by lowering the pH of the water matrix.

Most separator technologies will have an optimum range of oil droplet sizes that can be effectively treated.

Analysing the oily water to determine droplet size can be performed with a video particle analyser. Each separator technology will have its own performance curve outlining optimum performance based on oil droplet size. the most common separators are gravity tanks or pits, API oil-water separators or plate packs, chemical treatment via DAFs, centrifuges, media filters and hydrocyclones.

API Separators

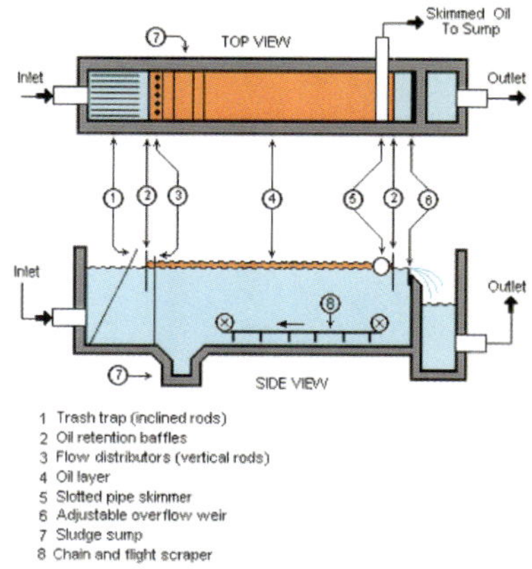

A typical API oil-water separator used in many industries

Many oils can be recovered from open water surfaces by skimming devices. Considered a dependable and cheap way to remove oil, grease and other hydrocarbons from water, oil skimmers can sometimes achieve the desired level of water purity. At other times, skimming is also a cost-efficient method to remove most of the oil before using membrane filters and chemical processes. Skimmers will prevent filters from blinding prematurely and keep chemical costs down because there is less oil to process.

Because grease skimming involves higher viscosity hydrocarbons, skimmers must be equipped with heaters powerful enough to keep grease fluid for discharge. If floating grease forms into solid clumps or mats, a spray bar, aerator or mechanical apparatus can be used to facilitate removal.

However, hydraulic oils and the majority of oils that have degraded to any extent will also have a soluble or emulsified component that will require further treatment to eliminate. Dissolving or emulsifying oil using surfactants or solvents usually exacerbates the problem rather than solving it, producing wastewater that is more difficult to treat.

The wastewaters from large-scale industries such as oil refineries, petrochemical plants, chemical plants, and natural gas processing plants commonly contain gross amounts of oil and suspended solids. Those industries use a device known as an API oil-water separator which is designed to separate the oil and suspended solids from their wastewater effluents. The name is derived from the fact that such separators are designed according to standards published by the American Petroleum Institute (API).

The API separator is a gravity separation device designed by using Stokes Law to define the rise velocity of oil droplets based on their density and size. The design is based on the specific gravitydifference between the oil and the wastewater because that difference is much smaller than the specific gravity difference between the suspended solids and water. The suspended solids settles to the bottom of the separator as a sediment layer, the oil rises to top of the separator and the cleansed wastewater is the middle layer between the oil layer and the solids.

Typically, the oil layer is skimmed off and subsequently re-processed or disposed of, and the bottom sediment layer is removed by a chain and flight scraper (or similar device) and a sludge pump. The water layer is sent to further treatment for additional removal of any residual oil and then to some type of biological treatment unit for removal of undesirable dissolved chemical compounds.

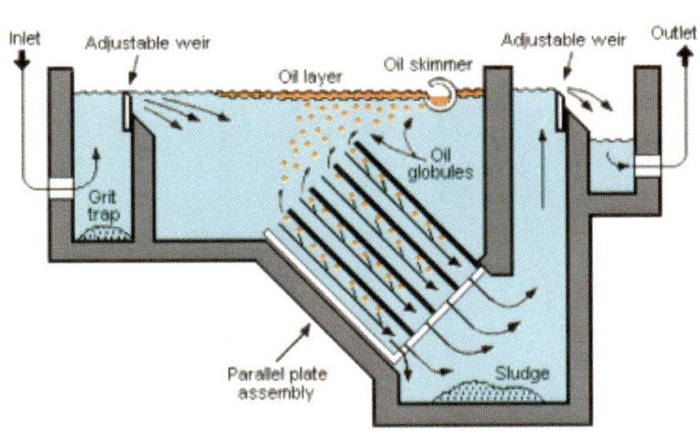

A typical parallel plate separator

Parallel plate separators are similar to API separators but they include tilted parallel plate assemblies (also known as parallel packs). The parallel plates provide more surface for suspended oil droplets to coalesce into larger globules. Such separators still depend upon the specific gravity between the suspended oil and the water. However, the parallel plates enhance the degree of oil-water separation. The result is that a parallel plate separator requires significantly less space than a conventional API separator to achieve the same degree of separation.

Hydrocyclone Oil Separators

Hydrocyclone oil separators operate on the process where wastewater enters the cyclone chamber and is spun under extreme centrifugal forces more than 1000 times the force of gravity. This force causes the water and oil droplets to separate. The separated oil is discharged from one end of the cyclone where treated water is discharged through the opposite end for further treatment, filtration or discharge.

Removal of Biodegradable Organics

Biodegradable organic material of plant or animal origin is usually possible to treat using extended conventional sewage treatment processes such as activated sludgeor trickling filter. Problems can arise if the wastewater is excessively diluted with washing water or is highly concentrated such as undiluted blood or milk. The presence of cleaning agents, disinfectants, pesticides, or antibiotics can have detrimental impacts on treatment processes.

Activated Sludge Process

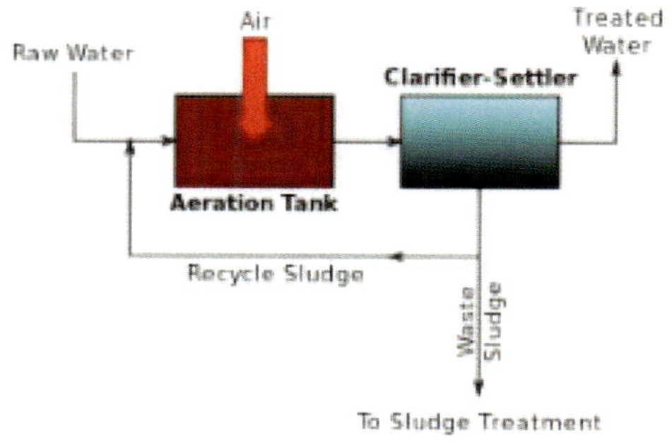

A generalized diagram of an activated sludge process

Activated sludge is a biochemical process for treating sewage and industrial wastewater that uses air (or oxygen) and microorganisms to biologically oxidize organic pollutants, producing a waste sludge (or floc) containing the oxidized material. In general, an activated sludge process includes:

An aeration tank where air (or oxygen) is injected and thoroughly mixed into the wastewater.

A settling tank (usually referred to as a clarifier or "settler") to allow the waste sludge to settle. Part of the waste sludge is recycled to the aeration tank and the remaining waste sludge is removed for further treatment and ultimate disposal.

Trickling Filter Process

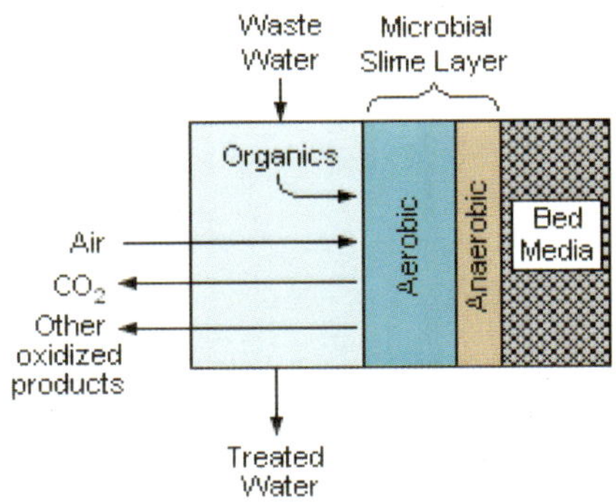

Figure: A schematic cross-section of the contact face of the bed media in a trickling filter

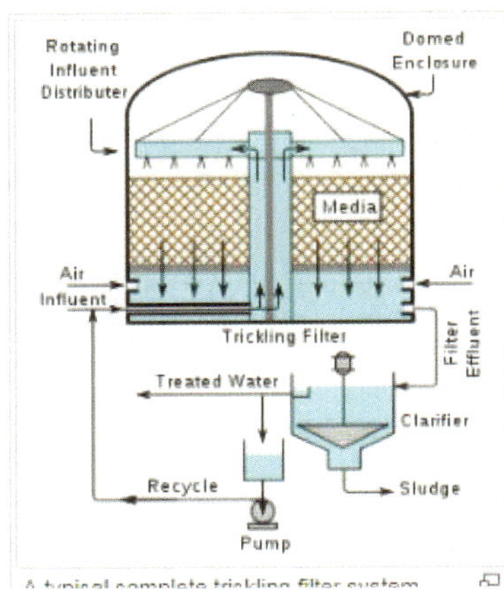

A typical complete trickling filter system

A trickling filter consists of a bed of rocks, gravel, slag, peat moss, or plastic media over which wastewater flows downward and contacts a layer (or film) of microbial slime covering the bed media. Aerobic conditions are maintained by forced air flowing through the bed or by natural convection of air. The process involves adsorption of organic compounds in the wastewater by the microbial slime layer, diffusion of air into the slime layer to provide the oxygen required for the biochemical oxidation of the organic compounds. The end products include carbon dioxide gas, water and other products of the oxidation. As the slime layer thickens, it becomes difficult for the air to penetrate the layer and an inner anaerobic layer is formed.

The fundamental components of a complete trickling filter system are:

- A bed of filter medium upon which a layer of microbial slime is promoted and developed.

- An enclosure or a container which houses the bed of filter medium.
- A system for distributing the flow of wastewater over the filter medium.
- A system for removing and disposing of any sludge from the treated effluent.

The treatment of sewage or other wastewater with trickling filters is among the oldest and most well characterized treatment technologies.

A trickling filter is also often called a trickle filter, trickling bio filter, bio filter, biological filter or biological trickling filter.

Removal of Other Organics

Synthetic organic materials including solvents, paints, pharmaceuticals, pesticides, products from coke production and so forth can be very difficult to treat. Treatment methods are often specific to the material being treated. Methods include advanced oxidation processing, distillation, adsorption, vitrification, incineration, chemical immobilization or landfill disposal. Some materials such as some detergents may be capable of biological degradation and in such cases; a modified form of wastewater treatment can be used.

Removal of Acids and Alkalis

Acids and alkalis can usually be neutralized under controlled conditions. Neutralization frequently produces a precipitate that will require treatment as a solid residue that may also be toxic. In some cases, gases may be evolved requiring treatment for the gas stream. Some other forms of treatment are usually required following neutralization.

Waste streams rich in hardness ions as from de-ionization processes can readily lose the hardness ions in a buildup of precipitated calcium and magnesium salts. This precipitation process can cause severe furring of pipes and can, in extreme cases, cause the blockage of disposal pipes. A 1-metre diameter industrial marine discharge pipe serving a major chemicals complex was blocked by such salts in the 1970s. Treatment is by concentration of de-ionization waste waters and disposal to landfill or by careful pH management of the released wastewater.

Removal of Toxic Materials

Toxic materials including many organic materials, metals (such as zinc, silver, cadmium, thallium, etc.) acids, alkalis, non-metallic elements (such as arsenic or selenium) are generally resistant to biological processes unless very dilute. Metals can often be precipitated out by changing the pH or by treatment with other chemicals. Many, however, are resistant to treatment or mitigation and may require concentration followed by landfilling or recycling. Dissolved organics can be incinerated within the wastewater by the advanced oxidation process.

Smart Capsules

Molecular encapsulation is a technology that has the potential to provide a system for the recyclable removal of lead and other ions from polluted sources. Nano, micro and milli capsules, with sizes in the ranges 10 nm-1μm,1μm-1mm and >1mm, respectively, are particles that have an active reagent (core) surrounded by a carrier (shell).There are three types of capsule under investigation: alginate-based capsules, carbon nanotubes, polymer swelling capsules. These capsules provide a possible means for the remediation of contaminated water.

Agricultural Waste

Agricultural waste refers to waste produced from agricultural operations, including waste from farms, poultry houses, and slaughterhouses. In addition to being potentially damaging to the environment, agricultural waste may expose workers to harmful biological material (biohazards). Storing agricultural waste can multiply the hazards associated with it since stored waste can release harmful gases.

Specific biohazards related to agricultural waste include:

- Fungi
- Molds
- Bacteria
- Viruses
- Animal-borne diseases (zoonosis)

Removal of waste products such as manure from animal and poultry houses can potentially expose workers to agricultural dusts, resulting in respiratory problems, such as occupational asthma. Exposure to waste-containing pesticides and other agrochemicals poses additional risks. Personal protective equipment (PPE) and personal respiratory equipment (PRE) can protect workers from many types of exposure.

Many agricultural wastes are economically productive resources. Agricultural slurry, for example, can be converted to fertilizer. This incentivizes waste storage as an economic activity. Safety hazards involved in waste storage include exposure to environments containing carbon dioxide (CO_2), ammonia, hydrogen sulfide, and methane gas. At unsafe levels, these gases can be toxic, flammable, and potentially explosive. Waste storage environments may also produce low-oxygen environments that can endanger workers who enter without proper protection.

Employers have a duty to provide employees with the equipment and training necessary for safely handling agricultural wastes. Workplaces are required to have first aid supplies available for treating injuries on-site and plans for addressing accidents or spills involving agricultural waste. Workplace risk is also reduced via requirements for a minimum safe distance between worker camps and environments containing agricultural waste.

References

- "Waste - Disposal of Laboratory Wastes (GUIDANCE) | Current Staff | University of St Andrews". www.st-andrews.ac.uk. Retrieved 2016-02-04
- Hazardous-waste-management, technology: britannica.com, Retrieved 17 April 2018
- Agricultural-waste-573: safeopedia.com, Retrieved 27 May 2018
- Maczulak, Anne Elizabeth (2010). Pollution: Treating Environmental Toxins. New York: Infobase Publishing. p. 120. ISBN 9781438126333
- Treatment-storage-and-disposal, hazardous-waste-management, technology: britannica.com, Retrieved 07 July 2018
- Laboratory, National Research Council (US) Committee on Prudent Practices in the (2011-01-01). "Management of Waste". Retrieved 10 March 2016
- Remedial-action, hazardous-waste-management, technology: britannica.com, Retrieved 29 June 2018

3

Environmental Impacts of Industrial Activities

The steep rise of industrialization has led to severe impact on the natural and man-made environments. An understanding of the environmental impact of industrial activities requires a study of the major industries of the 21st century namely the plastic industry, construction industry, oil industry, agricultural industry and mining industry. This chapter extensively discusses all such aspects for a comprehensive understanding.

Mining Industry

Mining industry is the branch of manufacturing and trade based on the extraction of ores, fossil fuels, minerals, stone, clay, gravel, and similar commodities. It is considered as one of the fastest growing economies all over the world but in particular, the contribution of mining to the country, especially those who export mineral products, are felt intensively.

Mining Industry Categories

Coal Mining Industry: All establishments primarily engaged in producing bituminous coal, anthracite, and lignite. This includes mining operations and preparation plants. This does not include the production of coal briquettes and packaged fuel nor does it include the production of gas and hydrocarbon liquids from coal at the mine site.

Metal Mining Industry: All establishments primarily engaged in mining, developing mines, or exploring for metallic minerals. These ores are valued chiefly for the metals contained, to be recovered for use as such or as constituents of alloys, chemicals, pigments or other products. This does not include the refinement of the ores.

Nonmetallic Mineral Mining Industry: All establishments primarily engaged in mining or quarrying, developing mines, or exploring for nonmetallic minerals, except fuels. This also includes certain well and brine operations, and primary preparation plants, such as those engaged in crushing, grinding, washing, or other concentration. This does not include establishments engaged in manufacturing cement or lime, engaged in cutting and finishing stone, or those engaged in manufacturing brick and other structural clay products.

Oil and Gas Extraction Industry: All establishments primarily engaged in producing crude petroleum and natural gas; extracting oil from sands and oil shale; producing natural gasoline and cycle condensate; and producing gas and hydrocarbon liquids from coal at the mine site. This includes activities such as exploration, drilling, oil and gas well operation and maintenance, the operation of natural gasoline and cycle plants, etc. This does not include pipeline transportation.

Types of Mining

There are basically two types of mining:

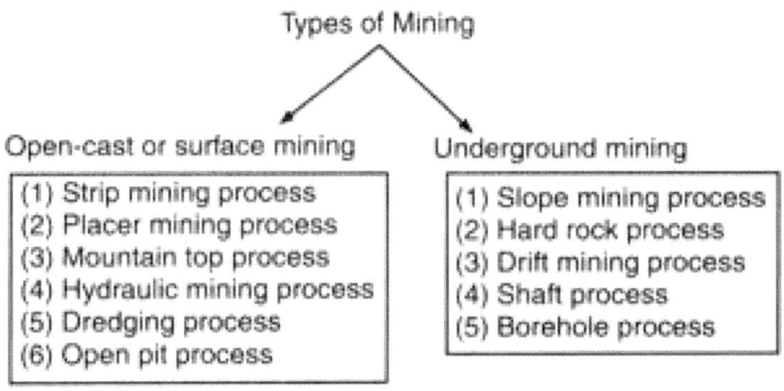

(i) Surface Mining

In this type of mining the minerals or rocks that are to be mined are exposed at the surface or very close to the surface. To extract the materials, earlier the miners used to dig an open-pit on the surface and then dig out the materials. But nowadays, giant earthmovers remove the top soil and the rocks and the material is extracted.

There are many types of surface mining processes. They are:

Strip Mining Process

As the name suggests the surface of the earth is stripped. In this process the soil and rocks that lie above is removed generally by heavy machinery and then the material is extracted.

This type is possible only when the targeted materials are relatively near the surface.

The mineral generally extracted is coal or some kinds of sedimentary rocks.

Placer Mining Process

In this process alluvial deposits in sand or gravel are extracted. This process does not involve the usage of any sort of heavy materials and can be considered relatively easy when compared to others. Generally gold and other gemstones are mined in this process.

Mountain Top Process

This process involves blasting of the top of mountains expose the coal below it. This process is relatively new.

Hydraulic Mining Process

In this process high pressure water jets are used to dislodge rocks and minerals. During earlier times gold was found in an easier way by this method. However, this process has been discontinued due to environmental concerns.

Dredging Process

It is the process of underwater excavation by deepening a water body. In this method sediments and other substances are removed from harbors, rivers and other water bodies and minerals are extracted.

Open Pit Process

The easiest and the cheapest way to mine materials that are close to the surface, large open holes are dug in the ground. Sometimes, explosives are used to get large blocks of rocks out of the way.

(ii) Underground Mining

Regarding underground mining there are five processes. They are:

Slope Mining Process

A type of underground mining, slopes are made into the ground and the desired material is accessed. This is done when the minerals are located far enough and surface mining cannot be employed to reach it. Generally, coal is mined in this way.

Hard Rock Process

Here tunnels deep in the ground are dug sometimes with dynamite or large drills. The tunnels are supported by pillars through which miners can move about. Tin, lead, copper, silver, gold etc. are mined with this process. This is typically the mine that comes to our minds when we think of it.

Drift Mining Process

This process is carried out when the material is situated sideways of a mountain. The materials are easier to access and the mouth is made slightly lower than the resource area so as to allow gravity to pull down the materials easily. Generally, coal or iron ore is mined through this process.

Shaft Process

The deepest form of underground mining, this is done by excavating a vertical passageway deep down. The materials to be extracted are situated deep inside and elevators are used to take the miners up and down. It is kept in mind that the tunnels are made airy for miners to work without any problems. Generally, coal is mined in this process.

Borehole Process:

Using a drill a deep hole is dug and a high pressure water-jet is used to force the materials up the hole. Sand, galena, gold etc. are mined by this process.

Impacts of Mining

Mining has the potential to have severely adverse effects on the environment including loss of biodiversity, erosion, contamination of surface water, ground water, and soil. The formation of sinkholes is also possible. Other than environmental damage, mining may also affect the surrounding population's health as a result of contamination caused by the leakage of chemicals.

In some countries, mining companies are expected to adhere to rehabilitation and environmental codes, making sure that the area mined is eventually transformed back into its original form, or at least as much as it is possible. Different types of mining methods can have significant public health and environmental effect.The erosion of exposed tailing dams, mine dumps, hillsides, and the resultant siltation of creeks, drainages, and rivers can affect the neighboring areas. A good example is Papua New Guinea's Ok Tedious Mine. Mining around farming areas may either destroy or disturb the crop lands or productive grazing lands while in wilderness areas it can cause either the disturbance or the destruction of ecosystems.

Water Pollution

Mining can have harmful effects on surrounding surface and groundwater if protective measures are not taken. The result can be unnaturally high concentrations of some chemicals, such as arsenic, sulfuric acid, and mercury over a significant area of surface or subsurface. Runoff of mere soil or rock debris -although non-toxic- also devastates the surrounding vegetation. The dumping of the runoff in surface waters or in forests is the worst option here. Submarine tailings disposal is regarded as a better option (if the soil is pumped to a great depth). Mere land storage and refilling of the mine after it has been depleted is even better, if no forests need to be cleared for the storage of the debris. There is potential for massive contamination of the area surrounding mines due to the various chemicals used in the mining process as well as the potentially damaging compounds and metals removed from the ground with the ore. Large amounts of water produced from mine drainage, mine cooling, aqueous extraction and other mining processes increases the potential for these chemicals to contaminate ground and surface water. In well-regulated mines, hydrologists and geologists take careful measurements of water and soil to exclude any type of water contamination that could be caused by the mine›s operations. The reducing or eliminating of environmental degradation is enforced in modern American mining by federal and state law, by restricting operators to meet standards for protecting surface and ground water from contamination. This is best done through the use of non-toxic extraction processes as bioleaching.

Acid mine drainage in Portugal

Acid Rock Drainage

Sub-surface mining often progresses below the water table, so water must be constantly pumped out of the mine in order to prevent flooding. When a mine is abandoned, the pumping ceases, and water floods the mine. This introduction of water is the initial step in most acid rock drainage situations.

Acid rock drainage occurs naturally within some environments as part of the rock weathering process but is exacerbated by large-scale earth disturbances characteristic of mining and other large construction activities, usually within rocks containing an abundance of sulfide minerals. Areas where the earth has been disturbed (e.g.

construction sites, subdivisions, and transportation corridors) may create acid rock drainage. In many localities, the liquid that drains from coal stocks, coal handling facilities, coal washeries, and coal waste tips can be highly acidic, and in such cases it is treated as acid mine drainage (AMD).

The same type of chemical reactions and processes may occur through the disturbance of acid sulfate soils formed under coastal or estuarine conditions after the last major sea level rise, and constitutes a similar environmental hazard.

The five principal technologies used to monitor and control water flow at mine sites are diversion systems, containment ponds, groundwater pumping systems, subsurface drainage systems, and subsurface barriers. In the case of AMD, contaminated water is generally pumped to a treatment facility that neutralizes the contaminants.

A 2006 review of environmental impact statements found that "water quality predictions made after considering the effects of mitigations largely underestimated actual impacts to groundwater, seeps, and surface water".

Heavy Metals

Dissolution and transport of metals and heavy metals by run-off and ground water is another example of environmental problems with mining, such as the Britannia Mine, a former copper mine near Vancouver, British Columbia. Tar Creek, an abandoned mining area in Picher, Oklahoma that is now an Environmental Protection Agency superfund site, also suffers from heavy metal contamination. Water in the mine containing dissolved heavy metals such as lead and cadmium leaked into local groundwater, contaminating it. Long-term storage of tailings and dust can lead to additional problems, as they can be easily blown off site by wind, as occurred at Skouriotissa, an abandoned copper mine in Cyprus.

Effects On Biodiversity

The implantation of a mine is a major habitat modification, and smaller perturbations occur on a larger scale than exploitation site, mine-waste residuals contamination of the environment for example. Adverse effects can be observed long after the end of the mine activity. Destruction or drastic modification of the original site and anthropogenic substances release can have major impact on biodiversity in the area. Destruction of the habitat is the main component of biodiversity losses, but direct poisoning caused by mine-extracted material, and indirect poisoning through food and water, can also affect animals, vegetation and microorganisms. Habitat modification such as pH and temperature modification disturb communities in the area. Endemic species are especially sensitive, since they need very specific environmental conditions. Destruction or slight modification of their habitat puts them at the risk of extinction. Habitats can be damaged when there is not enough terrestrial product as well as by non-chemicals products, such as large rocks from the mines that are discarded in the surrounding landscape with no concern for impacts on natural habitat.

The Ok Tedi River is contaminated by tailings from a nearby mine.

Concentrations of heavy metals are known to decrease with distance from the mine, and effects on biodiversity follow the same pattern. Impacts can vary greatly depending on mobility and bioavailability of the contaminant: less-mobile molecules will stay inert in the environment while highly mobile molecules will easily move into another compartment or be taken up by organisms. For example, speciation of metals in sediments could modify their bioavailability, and thus their toxicity for aquatic organisms.

Bio magnification plays an important role in polluted habitats: mining impacts on biodiversity should be, assuming that concentration levels are not high enough to directly kill exposed organisms, greater on the species on top of the food chain because of this phenomenon.

Adverse mining effects on biodiversity depend a great extent on the nature of the contaminant, the level of concentration at which it can be found in the environment, and the nature of the ecosystem itself. Some species are quite resistant to anthropogenic disturbances, while some others will completely disappear from the contaminated zone. Time alone does not seem to allow the habitat to recover completely from the contamination. Remediation takes time, and in most cases will not enable the recovery of the original diversity present before the mining activity took place.

Aquatic Organisms

The mining industry can impact aquatic biodiversity through different ways. Direct poisoning is the first one, and risks are higher when contaminants are mobile in the sediment or bioavailable in the water. Mine drainage can modify water pH, and it is hard to differentiate direct impact on organisms from impacts caused by pH changes. Effects can nonetheless be observed and proved to be caused by pH modifications. Contaminants can also affect aquatic organisms through physical effects: streams with high concentrations of suspended sediment limit light, thus diminishing algae biomass. Metal oxide deposition can limit biomass by coating algae or their substrate, thereby preventing colonization.

Contaminated Osisko lake in Rouyn-Noranda

Factors that impact communities in acid mine drainage sites vary temporarily and seasonally: temperature, rainfall, pH, salinisation and metal quantity all display variations on the long term, and can heavily affect communities. Changes in pH or temperature can affect metal solubility, and thereby the bioavailable quantity that directly impact organisms. Moreover, contamination persists over time: ninety years after a pyrite mine closure, water pH was still very low and microorganisms populations consisted mainly of acidophil bacteria.

Microorganisms

Algae communities are less diverse in acidic water containing high zinc concentration, and mine drainage stress decrease their primary production. Diatoms community is greatly modified by any chemical change. pH phytoplankton assemblage,and high metal concentration diminishes the abundance of planktonic species. Some diatom species may however grow in high-metal-concentration sediments. In sediments close to the surface, cysts suffer from corrosion and heavy coating. In very polluted conditions, total algae biomass is quite low, and the planktonic diatom community missing. In case of functional complementarity however, it is possible that phytoplankton and zooplankton mass remains stable.

Macroorganisms

Water insect and crustacean communities are modified around a mine, resulting in a low trophic completeness and community being dominated by predators.

However, biodiversity of macroinvertebrates can remain high, if sensitive species are replaced with tolerant ones. When diversity is on the contrary reduced, there is sometimes no effect of stream contamination on abundance or biomass, suggesting that tolerant species fulfilling the same function take the place of sensible species in polluted sites. pH diminution in addition to elevated metal concentration can also have adverse effects on macroinvertebrates' behaviour, showing that direct toxicity is not the only issue. Fishes are also affected by pH, temperature variations and chemical concentrations.

Terrestrial Organisms

Vegetation

Soils' texture and water content can be greatly modified in disturbed sites, leading to plants communities changes in the area. Most of the plants have a low concentration tolerance for metals in the soil, but sensitivity differs among species. Grass diversity and total cover is less affected by high contaminant concentration than forbs and shrubs. Mines waste-material rejects or traces due to mining activity can be found in the vicinity of the mine, sometimes pretty far away from the source. Established plants cannot move away from perturbations, and will eventually die if their habitat is contaminated by heavy metals or metalloids at concentration too elevated for their physiology. Some species are more resistant and will survive these levels, and some non-native species that can tolerate these concentrations in the soil, will migrate in the mine surrounding lands to occupy the ecological niche.

Plants can be affected through direct poisoning, for example arsenic soil content reduces bryophyte diversity. Soil acidification through pH diminution by chemical contamination can also lead to a diminished species number. Contaminants can modify or disturb microorganisms, thus modifying nutrient availability, causing a loss of vegetation in the area. Some tree roots avoid the deeper soil layer in order to avoid the contaminated zone, and thus miss anchorage and might be uprooted by the wind when their height and shoot weight increase. In general, root exploration is reduced in contaminated areas compared to non-polluted ones. Even in reclaimed habitats, plant species diversity is lower than in undisturbed areas.

Cultivated crops might be a problem near mines. Most crops can grow on weakly contaminated sites, but yield is generally lower than it would have been in regular growing conditions. Plants also tend to accumulate heavy metals in their aerian organs, possibly leading to human intake through fruits and vegetables. Regular consumptions might lead to health problems caused by long-term metal exposure. Cigarettes made from tobacco growing on contaminated sites might as well have adverse effects on human population, as tobacco tends to accumulate cadmium and zinc in its leaves.

Animals

Malartic mine - Osisko

Habitat destruction is one of the main issues of mining activity. Huge areas of natural habitat are destroyed during mine construction and exploitation, forcing animals to leave the site.

Animals can be poisoned directly by mine products and residuals. Bioaccumulation in the plants or the smaller organisms they eat can also lead to poisoning: horses, goats and sheep are exposed in certain areas to potentially toxic concentration of copper and lead in grass. There are fewer ant species in soil containing high copper levels, in the vicinity of a copper mine.If fewer ants are found, chances are great that other organisms leaving in the surrounding landscape are strongly affected as well by this high copper levels, since ants are a good environmental control: they live directly in the soil and are thus pretty sensitive to environmental disruptions.

Microorganisms

Because of their size, microorganisms are extremely sensitive to environmental modification,such as modified pH, temperature changes or chemicals concentration. For example, the presence of arsenic and antimony in soils led to a diminution in total soil bacteria. Moreover, as in water, a small change in the soil pH can provoke the remobilization of contaminants, in addition to the direct impact on pH-sensitive organisms.

Microorganisms have a wide variety of genes among their total population, so there is a greater chance of survival of the species due to the existence of resistance or tolerance genes in some colonies, as long as modifications are not too extreme. Nevertheless, survival in these conditions will imply a big loss of gene diversity, resulting in reduced potential adaptations to subsequent changes. The presence of few developed soil in heavy metal contaminated areas could be a sign of reduced activity by soils microfauna and microflora, indicating a reduced number of individuals or reduced activity. Twenty years after disturbance, even in rehabilitation area, microbial biomass is still greatly reduced compared to undisturbed habitat. Arbuscular mycorrhiza fungi are especially sensitive to the presence of chemicals, and the soil is sometimes so disturbed that they are no longer able to associate with root plants. Some fungi possess however contaminant accumulation capacity, soil cleaning capacity by changing the biodisponibility of contaminants, and can protect plants from damages caused by chemicals. Their presence in contaminated sites could prevent loss of biodiversity due to mine-waste contamination, or allow bioremediation, that is, the removal of undesired chemicals from contaminated soils. On the contrary, some microbes can deteriorate the environment: which mean elevated SO_4 in the water can also increase microbial production of hydrogen sulfide, a toxin for many aquatic plants and organisms.

Effects Of Mine Pollution On Humans

Humans are also affected by mining. There are many diseases that can come from the pollutants that are released into the air and water during the mining process. For example, during smelting operations enormous quantities of air pollutants, such as the suspended particulate matter, SO_x, arsenic particles and cadmium, are emitted. Metals are usually emitted into the air as particulates.

There are also many occupational health hazards. Most of the miners suffer from various respiratory and skin diseases. Miners working in different types of mines suffer from asbestosis, silicosis, or black lung disease.. Humans are also affected by the occurrence of landslides and floods.

Deforestation

With open cast mining the overburden, which may be covered in forest, must be removed before the mining can commence. Although the deforestation due to mining may be small compared to the total amount it may lead to species extinction if there is a high level of local endemism.

Benefits of Mining

Most people underestimate the importance of the role of this sector in global economies, but little do they know national governments of countries who export minerals rely heavily on this industry to boost their economic performance.

As a matter of fact, many developing African countries are dependent on the exploration and the exploitation of their mineral resources such as coal, gold, ore, iron, and many other mineral sectors which are deemed to open many opportunities for many economies.

So the importance of this sector is highlighted with the economic significance of mining, especially since the economic benefits of mining are apparent.

Economic Benefits of Mining

Similarly, the economic benefits of mining also highlight the economic significance of mining, given that both concepts are subsequent consequences of each other.

With the economic significance of this sector affected by the economic benefits of mining, an economist always considers which significant decisions of the sector can positively impact the economy and highlight the industry's importance, hence the need for learning what the economic benefits of mining are. Some of these benefits include:

1. Providing livelihood to local communities.
2. Generating income for the local government thereby directly contributing to its economy.
3. Promoting a more efficient use of energy, saving people money.
4. Promoting environmental awareness through rehabilitation programs resulting to more renewable sources than ever.

With these benefits being apparent, the contribution of this sector to the economy is sure to be significant.

Contribution of Mining to the Economy

The contribution of the **mining sector to the economic situation** of a country is more significant than most people think. The role of this sector in everyday life is to provide essential materials to help facilitate the assembly of new ones (such as in mobile phones), making it necessary for a country to practice safe mining to keep the other industries going.

This is the reason why mining is important to daily life, because, without this sector, many industries will be crippled, especially the manufacturing industry.

Importance of Mining in Important to Daily Life

The economic importance of this industry has been underestimated by many, but it has been proven that mining plays a crucial economic role to keep the economy going and for a country to experience growth.

The facts mentioned earlier all highlight the impacts of this sector on a country's wealth, and it is important to take note what these are for you to understand better why it is necessary for everyday life.

Coal Mining

Coal mining is the extraction of coal deposits from the surface of Earth and from underground.

Coal is the most abundant fossil fuel on Earth. Its predominant use has always been for producing heat energy. It was the basic energy source that fueled the Industrial Revolution of the 18th and 19th centuries, and the industrial growth of that era in turn supported the large-scale exploitation of coal deposits. Since the mid-20th century, coal has yielded its place to petroleum and natural gas as the principal energy supplier of the world. The mining of coal from surface and underground deposits today is a highly productive, mechanized operation.

Tavantolgoi coal mine, southern Gobi, Mongolia

There is archaeological evidence that coal was burned in funeral pyres during the Bronze Age, 3,000 to 4,000 years ago, in Wales. Aristotle mentions coal ("combustible bodies") in his Meteorologica, and his pupil Theophrastus also records its use. The Romans in Britain burned coal before AD 400; cinders have been found among the ruins of Roman villas and towns and along the Roman wall, especially in Northumberland, near the outcrop of coal seams. The Hopi Indians of what is now the southwestern United States mined coal by picking and scraping and used it for heating, cooking, and in ceremonial chambers as early as the 12th century AD; in the 14th century they used it industrially in pottery making. Marco Polo reports its use as widespread in 13th-century China. The Domesday Book (1086), which recorded everything of economic value in England, does not mention coal. London's first coal arrived by sea in 1228, from the areas of Fife and Northumberland, where lumps broken from submarine outcroppings and washed ashore by wave action were gathered by women and children. Thereafter, the name sea coal was applied to all bituminous coal in England. Later in the century, monks began to mine outcroppings in the north of England.

Developments In Mine Entry

Shafts

Except for the Chinese, who may have mined coal underground, all the early coal seams were worked from the surface, in fully exposed outcroppings. In the later middle Ages, however, exhaustion of outcrop coal in many places forced a change from surface to underground, or shaft, mining. Early shaft mines were little more than wells widened as much as miners dared in the face of danger of collapse. Shafts were sunk on high ground, with adits—near-horizontal tunnels—for drainage driven into the side of the hill. In England some shallow mine shafts were exhausted as early as the 14th century, making it necessary to go deeper and expand mining at the shaft bottoms. These remained small operations; a record of 1684 shows 70 mines near Bristol, employing 123 workers. Greater depth created many problems. First, water could no longer simply be drained away. Crude methods were devised to lift it to the surface. A bucket-and-chain device was first powered by men and later by horses; a continuous belt of circular plates was drawn up through a pipe. Windmills were used for pumps. But shafts had to be restricted to depths of 90 to 105 meters (300 to 350 feet) and a mining radius of 180 meters. It was not until 1710 that the water problem was eased by Thomas Newcomen's steam atmospheric engine, which supplied a cheap and reliable power source for a vertical reciprocating lift pump.

Hoisting

Raising the coal itself was another problem. Manpower, operating a windlass, was replaced by horsepower; and, as the shafts went deeper, more horses were added. At Whitehaven in 1801, coal was hoisted 180 meters by four horses at the rate of 42–44 metric tons (46–48 tons) in nine hours. The introduction of the steam engine to hoist coal was a major turning point for the industry. Small steam-powered windlasses were successfully tried out about 1770. About 1840 the first cage was used to hoist the loaded car; and from 1840 onward advances in coal-mining techniques were rapid.

Ventilation

The presence of noxious and flammable gases caused miners to recognize the critical importance of ventilation in coal mines from the earliest days. Natural ventilation was afforded by level drainage tunnels driven from the sloping surface to connect with the shaft. Surface stacks above the shaft increased the efficiency of ventilation; their use continued in small mines until the early 20th century. The most reliable method, before the introduction of fans, was the use of a furnace at the shaft bottom or on the surface. Despite the hazard of fire and explosion, there were still a large number of furnaces operating, at least in non gassy mines, in the early 20th century.

Open-flame illumination, however, was a much more common cause of explosions until the introduction of the Davy safety lamp (about 1815), in which the flame is enclosed in a double layer of wire gauze that prevents ignition of flammable gases in the air of the mine. Presence of strong air currents, however, made even the Davy lamp unsafe.

Rotary ventilating fans were introduced in mines in the 18th century. Originally of wood and powered by steam, they were improved throughout the 19th and 20th centuries by the introduction of steel blades, electric power, and aerodynamically efficient shapes for the blades.

From Manual to Mechanized Extraction

Conventional Mining

Early European miners wedged coal out of the seam or broke it loose with a pick. After explosives were introduced, it was still necessary to undercut the coal seam with hand tools. The advent of steam, compressed air, and electricity brought relief from this hard, dangerous work. In 1868, after almost 100 years of trial and error, a commercially successful revolving-wheel cutter for undercutting the coal seam was introduced in England. This first powered cutting tool was soon improved by introduction of compressed air as a power source in place of steam. Later, electricity was used. The long wall cutter was introduced in 1891. Originally driven by compressed air and later electrified, it could begin at one end of a long face (the vertical, exposed cross section of a seam of coal) and cut continuously to the other.

Development of Continuous Mining

The conventional mining techniques described above, made up of the cyclic operations of cutting, drilling, blasting, and loading, developed in association with room-and-pillar mining. The oldest of the basic underground methods, room-and-pillar mining grew naturally out of the need to recover more coal as mining operations became deeper and more expensive. During the late 1940s, conventional techniques began to be replaced by single machines, known as continuous miners that broke off the coal from the seam and transferred it back to the haulage system. The Joy Ripper (1948) was the first continuous miner applicable to the room-and-pillar method.

Origins of Longwall Mining

The other principal method of modern mining, long wall mining, had been introduced as early as the 17th century and had found general use by the 19th century, but it had long been less productive than room-and-pillar mining. This began to change in the 1940s, when a continuous system involving the "plow" was developed by Wilhelm Loebbe of Germany. Pulled across the face of the coal and guided by a pipe on the face side of a segmented conveyor, the plow carved a gash off the bottom of the seam. The conveyor snaked against the face behind the advancing plow to catch the coal that chipped off from above the gash. Substantially reducing the labor required at the coal face (except that needed to install roof support), the Loebbe system quickly became popular in Germany, France, and the Low Countries.

The plow itself had limited application in British mines, but the power-advanced segmented conveyor became a fundamental part of equipment there, and in 1952 a simple continuous machine called the shearer was introduced. Pulled along the face astride the conveyor, the shearer bore a series of disks fitted with picks on their perimeters and mounted on a shaft perpendicular to the face. The revolving disks cut a slice from the coal face as the machine was pulled along, and a plow behind the machine cleaned up any coal that dropped between the face and the conveyor.

Roof Support

The technique of supporting the roof by rock bolting became common in the late 1940s and did much to provide an unobstructed working area for room-and-pillar mining, but it was a laborious and slow operation that prevented long wall mining from realizing its potential. In the late 1950s, however, powered, self-advancing roof supports were introduced by the British. Individually or in groups, these supports, attached to the conveyor, could be hydraulically lowered, advanced, and reset against the roof, thus providing a prop-free area for equipment (between the coal face and the first row of jacks) and a canopied pathway for miners (between the first and second rows of jacks).

Haulage

Manual Labor to Electric Power

In the first shaft mines, coal was loaded into baskets that were carried on the backs of men or women or loaded on wooden sledges or trams that were then pushed or hauled through the main haulage roadway to the shaft bottom to be hung on hoisting ropes or chains. In drift and slope mines, the coal was brought directly to the surface by these and similar methods. Sledges were pulled first by men and later by animals, including mules, horses, oxen, and even dogs and goats.

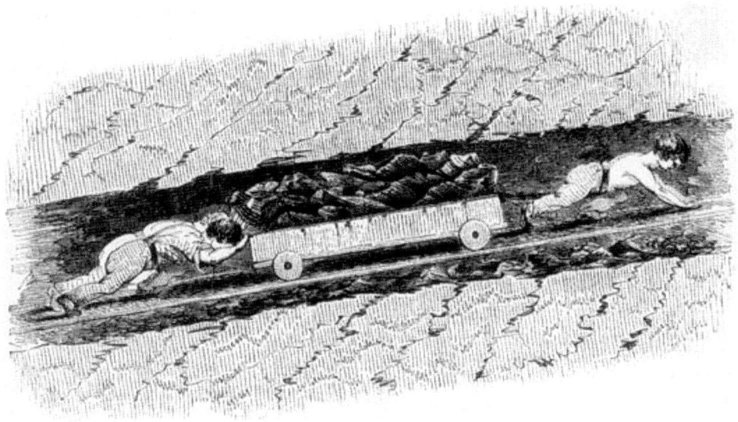

Children hauling coal up the slope of an English mine; from an engraving of the 1840s

Steam locomotives designed by Richard Trevithick were used in the fields of South Wales and Tyne and later in Pennsylvania and West Virginia, but they created too much smoke. Compressed-air locomotives, which appeared in the 1880s, proved expensive to operate. Electric locomotives, introduced in 1887, rapidly became popular, but mules and horses were still working in some mines as late as the 1940s.

Mechanized Loading

The loading by hand of broken coal into railcars was made obsolete early in the 20th century by mobile loaders. The Stanley Header, the first coal-loading machine used in the United States, was developed in England and tested in Colorado in 1888. Others were developed, but few progressed beyond the prototype stage until the Joy machine was introduced in 1914. Employing the gathering-arm principle, the Joy machine provided the pattern for future successful mobile loaders. After the introduction in 1938 of electric-powered, rubber-tired shuttle cars designed to carry coal from the loading machine to the elevator, mobile loading and haulage rapidly supplanted track haulage at the face of room-and-pillar mines.

Conveyors

In 1924 a conveyor belt was successfully used in an anthracite mine in central Pennsylvania to carry coal from a group of room conveyors to a string of cars at the mine entry. By the 1960s belts had almost completely replaced railcars for intermediate haulage.

Preparation

The history of coal preparation begins in the 19th century, with the adaptation of mineral-processing methods used for enriching metallic ores from their associated impurities. In the early years, larger pieces of coal were simply handpicked from pieces composed predominantly of mineral matter. Washing with mechanical devices to separate the coal from associated rocks on the basis of their density differences began during the 1840s.

At first, coal preparation was necessitated by the demand for higher heating values; another demand was for such special purposes as metallurgical coke for steelmaking. In recent years, as concern has grown over the emission of sulfur dioxide in the flue gases of power plants, coal preparation has taken on greater importance as a measure to remove atmospheric pollutants.

Coal Deposits

Coalification

In geologic terms, coal is a sedimentary rock containing a mixture of constituents, mostly of vegetal origin. Vegetal matter is composed mainly of carbon, hydrogen, oxygen, nitrogen, sulfur, and some inorganic mineral elements. When this material decays under water, in the absence of oxygen, the carbon content increases. The initial product of this decomposition process is known as peat. Peat can be formed in bogs, marshes, or freshwater swamps, and in fact huge freshwater swamps of the geologic past provided favorable conditions for the formation of thick peat deposits that over time became coal deposits. The transformation of peat to lignite is the result of pressure exerted by sedimentary materials that accumulate over the peat deposits. Even greater pressures and heat from movements of the Earth's crust (as occurs during mountain building), and occasionally from igneous intrusion, cause the transformation of lignite to bituminous and anthracite coal.

Major Coal Eras

Coal deposits are known to have formed more than 400 million years ago. Most anthracite and bituminous coals occur within the 299- to 359.2-million-year-old strata of the Carboniferous Period, the so-called first coal age. The formation of coal deposits continued through the Permian, Triassic, and Jurassic periods into the "second coal age," which includes the Cretaceous, Paleogene, and Neogene periods. Coals of the Cretaceous Period (145.5 million to 65.5 million years ago) are generally in the high-volatile to medium-volatile bituminous ranks. Cenozoic coals, formed less than 65.5 million years ago, are predominantly of the subbituminous and lignitic ranks.

Rank and Grade

The rank of a coal indicates the progressive changes in carbon, volatile matter, and probably ash and sulfur that take place as coalification progresses from the lower-rank lignite through the higher ranks of subbituminous, high-volatile bituminous, low-volatile bituminous, and anthracite. The rank of a coal should not be confused with its grade. A high rank (*e.g.*, anthracite) represents coal from a deposit that has undergone the greatest degree of devolatilization and contains very little mineral matter, ash, and moisture. On the other hand, any rank of coal, when cleaned of impurities through coal preparation, will be of a higher grade.

Resources and Reserves

Distribution Worldwide

Coal deposits are found in sedimentary rock basins, where they appear as successive layers, or seams, sandwiched between strata of sandstone and shale. There are more than 2,000 coal-bearing sedimentary basins distributed around the world. World coal resources—that is, the total amount of coal available in the world—are approximately 11 trillion tons. The distribution of the estimated coal resources of the world is approximately as follows: Europe (including Russia and the former Soviet republics) 49 percent; North America 29 percent; Asia 14 percent; Australia 6 percent; and Africa and South America 1 percent each. Distinct from coal resources are coal reserves, which are only those resources that are technically and economically minable at a particular time. The current recoverable coal reserves of the world are estimated at 760 billion tons. Their distribution by continent is: Europe 44 percent; North America 28 percent; Asia 17 percent; Australia 5 percent; Africa 5 percent; and South America 1 percent.

Economic Factors

Among the most important factors that influence the movement of a coal deposit from a resource to a reserve or vice versa are the price of coal in the energy market and the costs of producing the coal for that market. Currently, seams less than 30 centimeters (1 foot) in thickness are not considered economically recoverable. Furthermore, extraction from seams at great depth—*i.e.*, over 1,000 meters (3,300 feet)—presents great difficulties. Other geologic features, such as excessively steep seams, extensive faulting and folding, washouts created by erosion and sedimentation, and burnout of the coal seams by igneous intrusion, all affect the amount and quality of coal that can be recovered from a seam.

Prospecting and Exploration

The fundamental objective of coal prospecting is to discover coal resources through a search. In areas where coal mining has not been previously practiced, the search process should result in obtaining coal samples that give reasonable evidence of the existence of a coal seam. Once a seam has been discovered, considerable further work is necessary in order to advance knowledge of the particular geologic aspects and the extent of the coal deposit. The term coal exploration is used to describe these activities. Coal exploration includes activities and evaluations necessary to gather data for making decisions on such issues as the desirability of further exploration, the technical feasibility of mining (including favourable and unfavourable factors), and economic feasibility (including size of mine, coal quality assessment, marketability, and preparation of mined coal for market requirements).

Mapping

Geologic mapping is an important task in exploration. Mapping involves compiling detailed field notes on coal seams, strata above and below the seam, rock types, geologic structures, stream data, and man-made structures. Good maps and mapping techniques provide a means for planning and accomplishing exploration, development, reclamation, day-to-day operations, and equipment moves. Calculation of material volumes, location of physical elements, and determination of mining conditions are expedited by the use of maps. Maps also provide a method for recording data so that they can be organized and analyzed for ready reference.

Aerial photography and mapping methods (photogrammetry) are increasing in usefulness, particularly in the exploration and mining of surface deposits. Photogrammetric methods are relatively easy and inexpensive, can be adjusted to any scale, and are highly accurate in any terrain. Aerial photography can be conducted at an altitude designed to produce maps that show drainage configuration, roads, buildings, lakes, streams, timber, power lines, railroads, and fences or other features that may be missed by a ground survey.

Drilling

Drilling is the most reliable method of gathering information about a coal deposit and the mining conditions. It provides physical samples of the coal and overlying strata for chemical and physical analysis.

Spatial Patterns

Numerous factors are associated with a drilling program. One is the spatial pattern of the holes in an exploration area. When very large areas are being studied, hole spacings vary greatly and generally are not in any set pattern. When the program is narrowed to a specific target area, a grid pattern is most common. In areas where coal is known to exist, closely spaced drill-hole patterns are required.

Core Drilling and Rotary Drilling

A second factor associated with a drilling program is the choice between core drilling and rotary drilling. In core drilling, a hollow drill bit is attached to a core barrel so that cylindrical samples of the strata can be obtained. (Since the drill bit is faceted with diamonds for cutting the strata, this method is also called diamond core drilling.) Photographing the cores as they come out of the hole can provide data of great reliability.

In rotary drilling, the samples obtained are the chips and pulverized rock produced by the abrasive and chipping action of the drill bit. Rotary drilling is faster and comparatively less expensive than core drilling. In fact, it is not uncommon to drill down to the top of the coal seam by rotary drilling and then replace the drill tools for core drilling. In most programs, only 10 to 25 percent of the holes are actually cored for detailed information on overlying strata and coal. Coring of the coal seam itself, however, should closely approach 100 percent; if it does not, the analytical information obtained should be considered suspect.

Dozer Cutting

Exploration of coal outcrops may be accomplished with dozer cuts at regular intervals. Dozer cutting provides information on the attitude of the coal and on the nature of the overburden—important factors with regard to machine operation.

Geophysical Exploration

In geophysical exploration, the seismic, electric, magnetic, radiometric, and gravitational properties of earth materials are measured in order to detect anomalies that may be caused by the presence of mineral deposits. Their form of exploration may begin with airborne methods in regional and target-area investigations and continue with on-ground methods during detailed investigations. The most widely utilized airborne methods are, in increasing order of use, magnetic, magnetic plus radiometric, magnetic plus electromagnetic, and electromagnetic. These methods are almost always accompanied by aerial photography.

Ground geophysical methods have a major advantage over the airborne methods in that they are in direct contact with the earth. The principal methods are electrical, magnetic, electromagnetic, radiometric, gravimetric, and refraction-seismic. The drill-hole geophysical survey, called logging, is an important method of extending data acquisition beyond the drill hole. A combination of logging methods is advantageous: gamma-ray and density logging for identifying the type of coal present; gamma-ray (radiometric), resistivity (electric), and caliper logs for determining the thickness of the seam; and sonic and density logs for determining the condition of the roof and floor strata.

Environmental Impacts of Coal Mining

Land Use Management

Impact to Land and Surroundings

Strip mining severely alters the landscape, which reduces the value of the natural environment in the surrounding land. The land surface is dedicated to mining activities until it can be reshaped and reclaimed. If mining is allowed, resident human populations must be resettled off the mine site; economic activities, such as agriculture or hunting and gathering food and medicinal plants are interrupted. What becomes of the land surface after mining is determined by the manner in which the mining is conducted. Usually reclamation of disturbed lands to a land use condition is not equal to the original use. Existing land uses (such as livestock grazing, crop and timber production) are temporarily eliminated mining area. High-value, intensive-land-use areas like urban and transportation systems are not usually affected by mining operations. If mineral values are sufficient, these improvements may be removed to an adjacent area.

Strip mining eliminates existing vegetation, destroys the genetic soil profile, displaces or destroys wildlife and habitat, alters current land uses, and to some extent permanently changes the general topography of the area mined. Adverse impacts on geological features of human interest may occur in a coal strip mine. Geomorphic and geophysical features and outstanding scenic resources may be sacrificed by indiscriminate mining. Paleontological, cultural, and other historic values may be endangered due to the disruptive activities of blasting, ripping, and excavating coal. Stripping of overburden eliminates and destroys archeological and historic features, unless they are removed beforehand.

The removal of vegetative cover and activities associated with the construction of haul roads, stockpiling of topsoil, displacement of overburden and hauling of soil and coal increase the quantity of dust around mining operations. Dust degrades air quality in the immediate area, has an adverse impact on vegetative life, and constitutes health and safety hazards for mine workers and nearby residents.

Surface mining disrupts virtually all aesthetic elements of the landscape. Alteration of land forms often imposes unfamiliar and discontinuous configurations. New linear patterns appear as material is extracted and waste piles are developed. Different colors and textures are exposed as vegetative cover is removed and overburden dumped to the side. Dust, vibration, and diesel exhaust odors are created (affecting sight, sound, and smell). Residents of local communities often find such impacts disturbing or unpleasant. In case of mountaintop removal, tops are removed from mountains or hills to expose thick coal seams underneath. The soil and rock removed is deposited in nearby valleys, hollows and depressions, resulting in blocked (and contaminated) waterways.

Removal of soil and rock overburden covering the coal resource may cause burial and loss of topsoil, exposes parent material, and creates large infertile wastelands. Soil disturbance and associated compaction result in conditions conducive to erosion. Soil removal from the area to be surface-mined alters or destroys many natural soil characteristics, and reduces its biodiversity and productivity for agriculture. Soil structure may be disturbed by pulverization or aggregate breakdown.

Mine collapses (or mine subsidence) have the potential to produce major effects above ground, which are especially devastating in developed areas. German underground coal-mining (especially in North Rhine-Westphalia) has damaged thousands of houses, and the coal-mining industries have set aside large sums in funding for future subsidence damages as part of their insurance and state-subsidy schemes. In a particularly spectacular case in the German Saar region (another historical coal-mining area), a suspected mine collapse in 2008 created an earthquake measuring 4.0 on the Richter magnitude scale, causing some damage to houses. Previously, smaller earthquakes had become increasingly common and coal mining was temporarily suspended in the area.

In response to negative land effects of coal mining and the abundance of abandoned mines in the US the federal government enacted the Surface Mining Control and Reclamation Act of 1977, which requires reclamation plans for future coal mining sites. These plans must be approved by federal or state authorities before mining begins.

Water Management

Surface mining may impair groundwater in numerous ways: by drainage of usable water from shallow aquifers; lowering of water levels in adjacent areas and changes in flow direction within aquifers; contamination of usable aquifers below mining operations due to infiltration (percolation) of poor-quality mine water; and increased infiltration of precipitation on spoil piles. Where coal or carbonaceous shale is present, increased infiltration may result in: increased runoff of poor-quality water and erosion from spoil piles, recharge of poor-quality water to shallow groundwater aquifers and poor-quality water flow to nearby streams.

The contamination of both groundwater and nearby streams may be for long periods of time. Deterioration of stream quality results from acid mine drainage, toxic trace elements, high content of dissolved solids in mine drainage water, and increased sediment loads discharged to streams. When coal surfaces are exposed, pyrite comes in contact with water and air and forms sulfuric acid. As water drains from the mine, the acid moves into the waterways; as long as rain falls on the mine tailings the sulfuric-acid production continues, whether the mine is still operating or not. Also waste piles and coal storage piles can yield sediment to streams. Surface waters may be rendered unfit for agriculture, human consumption, bathing, or other household uses.

To anticipate these problems, water is monitored at coal mines. The five principal technologies used to control water flow at mine sites are: diversion systems, containment ponds, groundwater pumping systems, subsurface drainage systems and subsurface barriers.

River Water Pollution

Coal-fired boilers/power plants when using coal or lignite rich in limestone produces ash containing calcium oxide (CaO). CaO readily dissolves in water to form slaked lime/Ca(OH)2 and carried by rain water to rivers/irrigation water from the ash dump areas. Lime softening process precipitates Ca and Mg ions/removes temporary hardness in the water and also converts sodium bicarbonates in river water into sodium carbonate. Sodium carbonate (washing soda) further reacts with the remaining Ca and Mg in the water to remove / precipitate the total hardness. Also water-soluble sodium salts present in the ash enhance the sodium content in water further. Thus river water is converted into soft water by eliminating Ca and Mg ions and enhancing Na ions by coal-fired boilers. Soft water application in irrigation(-surface or ground water) converts the fertile soils into alkaline sodic soils. River water alkalinity and sodicity due to accumulation of salts in the remaining water after meeting various transpiration and evaporation losses, become acute when many coal-fired boilers and power stations are installed in a river basin. River water sodicity affects downstream cultivated river basins located in China, India, Egypt, Pakistan, west Asia, Australia, western US, etc.

Waste Management

Aerial photograph of Kingston Fossil Plant coal fly ash slurry spill site taken the day after the event

The burning of coal leaves substantial quantities of fly ash, which is usually stored in impoundment ponds. In the low-coal-content areas waste forms spoil tip. The U.S. EPA classified the 44 sites as potential hazards to communities (which means the waste sites could cause death and significant property damage if an event such as a storm, a terrorist attack or a structural failure caused a spill). The U.S. EPA estimated that about 300 dry landfills and wet storage ponds are used around the country to store ash from coal-fired power plants. The storage facilities hold the noncombustible ingredients of coal and the ash trapped by equipment designed to reduce air pollution.

Wildlife

Surface mining of coal causes direct and indirect damage to wildlife. The impact on wildlife stems primarily from disturbing, removing and redistributing the land surface. Some impacts are short-term, and confined to the mine site however others have far-reaching, long-term effects.

The most direct effect on wildlife is destruction or displacement of species in areas of excavation and spoil piling. Pit and spoil areas are not capable of providing food and cover for most species of wildlife. Mobile wildlife species like game animals, birds, and predators leave these areas. More sedentary animals like invertebrates, reptiles, burrowing rodents and small mammals may be destroyed. The community of microorganisms and nutrient-cycling processes are upset by movement, storage, and redistribution of soil.

Degradation of aquatic habitats is a major impact by surface mining, and may be apparent many miles from a mining site. Sediment contamination of surface water is common with surface mining. Sediment yields may increase a thousand times their former level as a result of strip mining.

The effects of sediment on aquatic wildlife vary with the species and the amount of contamination. High sediment levels can kill fish directly, bury spawning beds, reduce light transmission, alter temperature gradients, fill in pools, spread streamflows over wider, shallower areas, and reduce production of aquatic organisms used as food by other species. These changes destroy the habitat of valued species, and may enhance habitat for less-desirable species. Existing conditions are already marginal for some freshwater fish in the United States, and the sedimentation of their habitat may result in their extinction. The heaviest sediment pollution of a drainage normally comes within 5 to 25 years after mining. In some areas, unvegetated spoil piles continue to erode even 50 to 65 years after mining.

The presence of acid-forming materials exposed as a result of surface mining can affect wildlife by eliminating habitat and by causing direct destruction of some species. Lesser concentrations can suppress productivity, growth rate and reproduction of many aquatic species. Acids, dilute concentrations of heavy metals, and high alkalinity can cause severe damage to wildlife in some areas. The duration of acidic-waste pollution can be long; estimates of the time required to leach exposed acidic materials in the Eastern United States range from 800 to 3,000 years.

Air Pollution

Air Emissions

> "In northern China, air pollution from the burning of fossil fuels, principally coal, is causing people to die on average 5.5 years sooner than they otherwise might."
>
> — *Tim Flannery,* Atmosphere of Hope

Coal and coal waste products (including fly ash, bottom ash and boiler slag) release approximately 20 toxic-release chemicals, including arsenic, lead, mercury, nickel, vanadium, beryllium, cadmium, barium, chromium, copper, molybdenum, zinc, selenium and radium, which are dangerous if released into the environment. While these substances are trace impurities, enough coal is burned that significant amounts of these substances are released.

During combustion, the reaction between coal and the air produces oxides of carbon, including carbon dioxide (CO_2, an important greenhouse gas), oxides of sulfur (mainly sulfur dioxide, SO_2), and various oxides of nitrogen (NO_x). Because of the hydrogenous and nitrogenous components of coal, hydrides and nitrides of carbon and sulfur are also produced during the combustion of coal in air. These include hydrogen cyanide (HCN), sulfur nitrate (SNO_3) and other toxic substances.

SO_2 and nitrogen oxide react in the atmosphere to form fine particles and ground-level ozone and are transported long distances, making it difficult for other states to achieve healthy levels of pollution control.

The wet cooling towers used in coal-fired power stations, etc. emit drift and fog which are also an environmental concern. The drift contains Respirable suspended particulate matter. In case of cooling towers with sea water makeup, sodium salts are deposited on nearby lands which would convert the land into alkali soil, reducing the fertility of vegetative lands and also cause corrosion of nearby structures.

Fires sometimes occur in coal beds underground. When coal beds are exposed, the fire risk is increased. Weathered coal can also increase ground temperatures if it is left on the surface. Almost all fires in solid coal are ignited by surface fires caused by people or lightning. Spontaneous combustion is caused when coal oxidizes and airflow is insufficient to dissipate heat; this more commonly occurs in stockpiles and waste piles, rarely in bedded coal underground. Where coal fires occur, there is attendant air pollution from emission of smoke and noxious fumes into the atmosphere. Coal seam fires may burn underground for decades, threatening destruction of forests, homes, roadways and other valuable infrastructure. The best-known coal-seam fire may be the one which led to the permanent evacuation of Centralia, Pennsylvania, United States.

Approximately 75 Tg/S per year of Sulfur Dioxide (SO_2) is released from burning coal. After release, the Sulfur Dioxide is oxidized to gaseous H_2SO_2 which scatters solar radiation, hence their increase in the atmosphere exerts a cooling effect on climate that masks some of the warming caused by increased greenhouse gases. Release of SO_2 also contributes to the widespread acidification of ecosystems.

Mercury Emissions

Power plants are responsible for half of the mercury emissions in the United States.

In New York State winds deposit mercury from the coal-fired power plants of the Midwest, contaminating the waters of the Catskill Mountains. Mercury is concentrated up the food chain, as it is converted into methyl mercury, a toxic compound which harms both wildlife and people who consume freshwater fish. The mercury is consumed by worms, which are eaten by fish, which are eaten by birds (including bald eagles). As of 2008, mercury levels in bald eagles in the Catskills had reached new heights. "People are exposed to methyl mercury almost entirely by eating contaminated fish and wildlife that are at the top of aquatic food chains." Ocean fish account for the majority of human exposure to methyl mercury; the full range of sources of methyl mercury in ocean fish is not well understood.

In February 2012, the U.S. EPA issued Mercury and Air Toxics Standards (MATS), which require all coal plants to substantially reduce mercury emissions. "Today [2011], more than half of all coal-fired power plants already deploy pollution control technologies that will help them meet these achievable standards. Once final, these standards will level the playing field by ensuring the remaining plants – about 40 percent of all coal-fired power plants – take similar steps to decrease dangerous pollutants."

Annual Excess Mortality And Morbidity

In 2008 the World Health Organization (WHO) and other organizations calculated that coal particulates pollution cause approximately one million deaths annually across the world, which is approximately one third of all premature deaths related to all air pollution sources, for example in Istanbul by lung diseases and cancer.

Pollutants emitted by burning coal include fine particulates ($PM_{2.5}$) and ground level ozone. Every year, the burning of coal without the use of available pollution control technology causes thousands of preventable deaths in the United States. A study commissioned by the Maryland nurses association in 2006 found that emissions from just six of Maryland's coal-burning plants caused 700 deaths per year nationwide, including 100 in Maryland. Since installation of pollution abatement equipment on one of these six, the Brandon Shores plant, now "produces 90 percent less nitrogen oxide, an ingredient of smog; 95 percent less sulfur, which causes acid rain; and vastly lower fractions of other pollutants."

Economic Costs

A 2001 EU-funded study known as ExternE, or Externalities of Energy, over the decade from 1995 to 2005 found that the cost of producing electricity from coal would double over its present value, if external costs were taken into account. These external costs include damage to the environment and to human health from airborne particulate matter, nitrogen oxides, chromium VI and arsenic emissions produced by coal. It was estimated that external, downstream, fossil fuel costs amount up to 1–2% of the EU's entire Gross Domestic Product (GDP), with coal being the main fossil fuel accountable, and this was before the external cost of global warming from these sources was even included. The study found that environmental and health costs of coal alone were €0.06/kWh, or 6 cents/kWh, with the energy sources of the lowest external costs being nuclear power €0.0019/kWh, and wind power at €0.0009/kWh.

High rates of motherboard failures in China and India appear to be due to "sulfurous air pollution produced by coal that's burned to generate electricity. It corrodes the copper circuitry," according to Intel researchers.

Greenhouse Gas Emissions

The combustion of coal is the largest contributor to the human-made increase of CO_2 in the atmosphere. Electric generation using coal burning produces approximately twice the greenhouse gasses per kilowatt compared to generation using natural gas.

Coal mining releases methane, a potent greenhouse gas. Methane is the naturally occurring product of the decay of organic matter as coal deposits are formed with increasing depths of burial, rising temperatures, and rising pressure over geological time. A portion of the methane produced is absorbed by the coal and later released from the coal seam (and surrounding disturbed strata) during the mining process. Methane accounts for 10.5 percent of greenhouse-gas emissions created through human activity. According to the Intergovernmental Panel on Climate Change, methane has a global warming potential 21 times greater than that of carbon dioxide over a 100-year timeline. The process of mining can release pockets of methane. These gases may pose a threat to coal miners, as well as a source of air pollution. This is due to the relaxation of pressure and fracturing of the strata during mining activity, which gives rise to safety concerns for the coal miners if not managed properly. The buildup of pressure in the strata can lead to explosions during (or after) the mining process if prevention methods, such as "methane draining", are not taken.

In 2008 James E. Hansen and Pushker Kharecha published a peer-reviewed scientific study analyzing the effect of a coal phase-out on atmospheric CO_2 levels. Their baseline mitigation scenario was a phase-out of global coal emissions by 2050. Under the Business as Usual scenario, atmospheric CO_2 peaks at 563 parts per million (ppm) in the year 2100. Under the four coal phase-out scenarios, atmospheric CO2 peaks at 422–446 ppm between 2045 and 2060 and declines thereafter.

Radiation Exposure

Coal also contains low levels of uranium, thorium, and other naturally occurring radioactive isotopes which, if released into the environment, may lead to radioactive contamination. Coal plants emit radiation in the form of radioactive fly ash, which is inhaled and ingested by neighbors, and incorporated into crops. A 1978 paper from Oak Ridge National Laboratory estimated that coal-fired power plants of that time may contribute a whole-body committed dose of 19 μSv/a to their immediate neighbors in a 500 m radius. The United Nations Scientific Committee on the Effects of Atomic Radiation's 1988 report estimated the committed dose 1 km away to be 20 μSv/a for older plants or 1 μSv/a for newer plants with improved fly ash capture, but was unable to confirm these numbers by test.

Excluding contained waste and unintentional releases from nuclear plants, coal-plants carry more radioactive wastes into the environment than nuclear plants per unit of produced energy. Plant-emitted radiation carried by coal-derived fly ash delivers 100 times more radiation to the surrounding environment than does the normal operation of a similarly productive nuclear plant. This comparison does not consider the rest of the fuel cycle, i.e., coal and uranium mining and refining and waste disposal. The operation of a 1000-MWe coal-fired power plant results in a nuclear radiation dose of 490 person-rem/year, compared to 136 person-rem/year, for an equivalent nuclear power plant including uranium mining, reactor operation and waste disposal.

Dangers To Miners

Historically, coal mining has been a very dangerous activity, and the list of historical coal mining disasters is long. The principal hazards are mine wall failures and vehicle collisions; underground mining hazards include suffocation, gas poisoning, roof collapse and gas explosions. Chronic lung diseases, such as pneumoconiosis(black lung) were once common in miners, leading to reduced life expectancy. In some mining countries black lung is still common, with 4,000 new cases of black lung every year in the US (4 percent of workers annually) and 10,000 new cases every year in China (0.2 percent of workers). Rates may be higher than reported in some regions.

In the United States, an average of 23 coal miners per year died in the decade 2007–2016. Recent U.S. coal-mining disasters include the Sago Mine disaster of January 2006. In 2007, a mine accident in Utah's Crandall Canyon Mine killed nine miners, with six entombed. The Upper Big Branch Mine disaster in West Virginia killed 29 miners in April 2010.

However, in lesser developed countries and some developing countries, many miners continue to die annually, either through direct accidents in coal mines or through adverse health consequences from working under poor conditions. China, in particular, has the highest number of coal mining related deaths in the world, with official statistics claiming that 6,027 deaths in 2004. To compare, 28 deaths were reported in the US in the same year. Coal production in China is twice that in the US,while the number of coal miners is around 50 times that of the US, making deaths in coal mines in China 4 times as common per worker (108 times as common per unit output) as in the US.

The Farmington coal mine disasterkills 78. West Virginia, US

Build-ups of a hazardous gas are known as damps:

- Black damp: a miture of carbon dioxide and nitrogen in a mine can cause suffocation. The anoxic condition results of depletion of oxygen in enclosed spaces, e.g. by corrosion.
- After damp: similar to black damp, after damp consists of carbon monoxide, carbon dioxide and nitrogen and forms after a mine explosion.
- Fire damp: consists of mostly methane, a highly flammable gas that explodes between 5% and 15% – at 25% it causes asphyxiation.
- Stink damp: so named for the rotten egg smell of the hydrogen sulphide gas, stink damp can explode and is also very toxic.

White damp: air containing carbon monoxide which is toxic, even at low concentrations

Firedamp explosions can trigger the much more dangerous coal dust explosions, which can engulf an entire pit. Most of these risks can be greatly reduced in modern mines, and multiple fatality incidents are now rare in some parts of the developed world. Modern mining in the US results in approximately 30 deaths per year due to mine accidents.

Auxiliary and Unit Operations

Those activities which are essential to maintain safe and productive operating conditions both at the working faces and in all parts of the mine are known as auxiliary operations. These include ground control, ventilation, haulage, drainage, power supply, lighting, and communications. Those activities which are conducted sequentially in a production cycle—*i.e.,* cutting and hauling the coal and supporting the immediate exposed roof after coal removal—are called unit operations. Unit operations are planned and conducted so as to use the auxiliary services most effectively for maintaining health and safety as well as productivity at the locations where coal is actually being mined.

Access

Accesses to a coal seam, called portals, are the first to be completed and generally the last to be sealed. A large coal mine will have several portals. Their locations and the types of facilities installed in them depend on their principal use, whether for worker and material transport, ventilation, drainage and power lines, or emergency services. In many cases, the surface facilities near a portal include bathhouses and a lamp room; coal handling, storage, preparation, and load-out facilities; a fan house; water- and waste-handling systems; maintenance warehouses; office buildings; and parking lots.

There are three types of portal: drift, slope, and shaft. Where a coal seam outcrops to the surface, it is common to drive horizontal entries, called drifts, into the coal seam from the outcrop. Where the coal seam does not outcrop but is not far below the surface, it is accessed by driving sloping tunnels through the intervening ground. Slopes are driven at as steep an angle as is practicable for transporting coal by belt. Commonly, a pair of slopes is driven (or a slope is divided into two separate airtight compartments) or ventilation and material transport. Where the minimum coal-seam depth exceeds 250 to 300 meters, it is common to drive vertical shafts. Poor ground conditions are another factor in selecting a shaft over a slope. Shafts, too, may be split into separate compartments for fresh air, return air, worker and supply transport, and coal haulage.

Capital and operating costs for coal haulage are lowest in a drift access. Capital investment for coal haulage in a shaft or a slope is somewhat similar, but operating costs are generally higher in a shaft, owing to the non-continuous nature of shaft coal-handling facilities. It has been estimated that shafts and slopes, drifts, and permanent equipment in these access openings may account for more than 30 percent of the capital investment in a large mine.

Ground Control And Roof Support

Overall ground control—*i.e.,* long-term stability of mine accesses and entries and subsidence control—can be regarded as an auxiliary operation, whereas supporting the roof at production faces (roof control) is a unit operation. Ground control is concerned with the design of underground entries, their widths, the distance between the entries, and the number of entries that can be driven as a set. A hierarchy of entries exists in underground coal mines. Main entries are driven so as to divide the property into major areas; they usually serve the life of the mine for ventilation and for worker and material transport. Submain entries can be regarded as feeders from the mains that subdivide each major area. From the submains, panel entries take off to subdivide further a block of coal into panels for orderly coal extraction.

In some cases, complete collapse of the overlying strata during extraction eventually travels to the surface, resulting in surface depressions. This effect is called subsidence. Clearly, the wider and more numerous the entries, the more effective they will be for ventilation, materials handling, and first-mining extraction percentage. However, with increased width may come problems in entry and pillar stability. Often, by limiting the first mining to a small fraction of the coal seam and by laying out large undisturbed blocks of coal, subsidence may be reduced. The science of rock mechanics is well advanced and is useful for understanding such stability problems and for the design of mine openings, pillar sizing, extraction techniques, and planned subsidence.

Roof support at the face (the area where coal is actively mined) is intended to hold the immediate roof above the coal face. In modern mechanized mines, roof bolting is the most common method employed. Steel bolts, usually 1.2 to 2 meters long and 15 to 25 millimeters in diameter, are inserted in holes drilled into the roof by an electric rotary drill and are secured by either friction or resin. The bolts are set in rows across the entry, 1.2 to 1.8 meters apart. Several theories explain how roof bolts hold the roof. These include the beam theory (roof bolts tie together several weak strata into one), the suspension theory (weak members of the strata are suspended from a strong anchor horizon), and the keying-effect theory (roof bolts act much like the keystone in an arch).

Inserting steel bolts to support the roof of an underground mine in West Virginia

Additional supporting systems for entries (mains, sub mains, and panels) include temporary or permanent hydraulic or friction props, cribs (made of timber or reinforced concrete block), yieldable steel arches, and roof trusses.

Haulage

Coal haulage, the transport of mined coal from working faces to the surface, is a major factor in underground-mine efficiency. It can be considered in three stages: face or section haulage, which transfers the coal from the active working faces; intermediate or panel haulage, which transfers the coal onto the primary or main haulage; and the main haulage system, which removes the coal from the mine. The fundamental difference between face, intermediate, and main haulages is that the last two are essentially auxiliary operations in support of the first. Face haulage systems must be designed to handle large, instantaneous production from the cutting machines, whereas the outer haulage systems must be designed to accommodate such surges from several operating faces. Use of higher-capacity equipment in combination with bins or bunkers is common. In addition, face haulage systems generally discharge onto ratio-feeders or feeder-breakers in order to even out the flow of material onto the intermediate systems and to break very large lumps of coal or rock to below a maximum size.

In room-and-pillar systems, electric-powered, rubber-tired vehicles called shuttle cars haul coal from the face to the intermediate haulage system. In some semi mechanized or manual long wall operations, chain haulage is used, while the face haulage equipment of choice in modern mechanized long wall systems is an armored face conveyor (AFC). In addition to carrying coal from the face, the AFC serves as the guide for the long wall shearer, which rides on it.

Intermediate haulage in coal mines is provided by panel belts or by mine cars driven by locomotives. Panel belts have widths ranging from 90 to 150 centimeters, the wider belts being used with long wall panels. The use of mine cars and locomotives requires detailed considerations of shuttle-car dumping ramps, locomotive switching requirements, the inventory of mine cars, and track layout for empties and loads. Locomotives are electric- or diesel-powered. Mainline haulage is also provided by belt or railcar. The major differences are only in the size, scope, and permanence of installations. For example, mainline belts are laid for the life of the mine and are much wider and faster than intermediate belts. Mainline locomotives are also much larger than intermediate locomotives, and mainline tracks are built to more exacting standards of speed and reliability.

For the transport of maintenance and operating supplies to the working sections, advantage is taken of the mainline, intermediate, and face haulage systems. Monorail systems or endless-rope haulage systems, which are much like ski lifts, are commonly used in intermediate and face systems to transport supplies to the working faces. In all-belt mines, it is not unusual to have trolley rail haulage for carrying workers and materials to and from the working face. Other supply haulage equipment includes scoops and battery- or diesel-powered trucks.

Ventilation

The primary purpose of underground-mine ventilation is to provide oxygen to the miners and to dilute, render harmless, and carry away dangerous accumulations of gases and dust. In some of the gassiest mines, more than six tons of air are circulated through the mine for every ton of coal mined. Air circulation is achieved by creating a pressure difference between the mine workings and the surface through the use of fans. Fresh air is conducted through a set of mine entries (called intakes) to all places where miners may be working. After passing through the workings, this air (now termed return air) is conducted back to the surface through another set of entries (called returns). The intake and return airstreams are kept separate. Miners generally work in the intake airstream, although occasionally work must be done in the return airways.

The task of bringing fresh air near the production faces is an important auxiliary operation, while the task of carrying this air up to the working faces—the locations of which may change several times in a shift—is the unit operation known as face ventilation. The major difference between main ventilation and face ventilation is the number and nature of the ventilation control devices (fans, stoppings, doors, regulators, and air-crossings). In face ventilation, plastic or plastic-coated nylon cloth is generally used to construct stoppings and to divide the air along a face into the two streams of intake and return air. Furthermore, the stoppings, which are generally hung from the roof, are not secured at the bottom, in case machinery and coal must be transported from one side to the other. Main ventilation stoppings and air crossings, on the other hand, are constructed of brick or blocks and coated with mortar; the fans, regulators, and doors are also of substantial construction.

Monitoring And Control

Advancements in sensor technology and in computer hardware and software capabilities are finding increasing application in underground coal mines, especially in the monitoring and control of ventilation, haulage, and machine condition. Long wall shearers and shields can be remotely operated, and continuous miners have also been equipped with automatic controls. The atmospheric environment is remotely monitored for air velocity, concentrations of various gases, and airborne dust; fans and pumps are also monitored continuously for their operational status and characteristics.

Health, Safety, And Environment

In coal mining—particularly underground coal mining—there are numerous conditions that can threaten the health and safety of the miners. For this reason, coal mining worldwide is heavily regulated through health and safety laws. Through the development of new equipment for personnel protection, new approaches to mine design, more effective emergency preparedness plans and procedures, and major changes in legislation, regulation, and enforcement, higher standards of health and safety are now achieved. For example, the self-contained self-rescuer (SCSR) represents a significant development in raising a miner's chances of survival and escape after an explosion, fire, or similar emergency contaminates the mine atmosphere with toxic gases. This lightweight, belt-wearable device is available worldwide and is mandated in several countries to be carried on the person whenever underground.

The effects of mining on the water, air, and land outside the mine are as important as those that occur in the mine. These effects may be felt both on- and off-site; in addition, they may vary in severity from simple annoyance and property damage to possibly tragic illness and death. Even abandoned lands from past mining activities present such problems as mine fires, precipitous slopes, waste piles, subsidence, water pollution, derelict land, and other hazards endangering general welfare and public health. Growing environmental consciousness has brought about a greater consideration of environmental factors in the planning, designing, and operating of mines.

Plastic Industry

The manufacture of plastics and other polymer materials is under the functional domain of the plastic industry. It is a part of the chemical industry and the petrochemical industry. It can be a major driving resource for the industrial sector, from the industrial areas of building and construction, packaging and transportation to aerospace and electronics. The use of engineering plastic, as raw material, is significant for it has better thermal and mechanical properties than ordinary plastic.

The plastics industry is divided into two major sectors, the inter-relationship of which can be seen in figure below. The first sector comprises the raw material suppliers who manufacture polymers and moulding compounds from intermediates which they may also have produced themselves. In terms of invested capital this is usually the largest of the two sectors. The second sector is made up of processors who convert the raw materials into saleable items using various processes such as extrusion and injection moulding. Other sectors include machinery manufacturers who supply equipment to the processors and suppliers of special additives for use within the industry.

Production Sequence in the Processing of Plastics

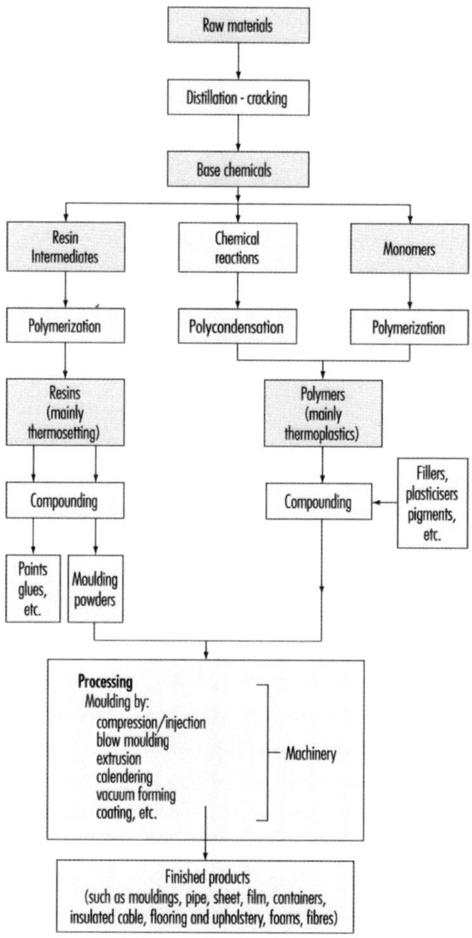

Polymer Manufacturing

Plastics materials fall broadly into two distinct categories: thermoplastics materials, which can be softened repeatedly by the application of heat and thermosetting materi-als, which undergo a chemical change when heated and shaped and cannot thereafter be reshaped by the application of heat. Several hundred individual polymers can be made with widely differing properties but as few as 20 types constitute about 90% of total world output. Thermoplastics are the largest group and their production is increasing at a higher rate than the thermosetting. In terms of production quantity the most important thermoplastics are high and low density polyethylene and polypropylene (the polyolefins), polyvinyl chloride (PVC) and polystyrene.

Important thermosetting resins are phenol-formaldehyde and urea-formaldehyde, both in the form of resins and moulding powders. Epoxy resins, unsaturated polyesters and polyurethanes are also significant. A smaller volume of "engineering plastics", for example, polyacetals, polyamides and polycarbonates, have a high value in use in critical applications.

The considerable expansion of the plastics industry in the post Second World War world was greatly facilitated by the broadening of the range of the basic raw materials feeding it; availability and price of raw materials are crucial to any rapidly developing industry. Traditional raw materials could not have provided chemical intermediates in sufficient quantities at an acceptable cost to facilitate the economic commercial production of large-tonnage plastics materials and it was the development of the petrochemicals industry which made growth possible. Petroleum as a raw material is abundantly available, easily transported and handled and was, until the oil crisis of the 1970s, relatively cheap. Therefore, throughout the world, the plastics industry is primarily tied to the use of intermediates obtained from oil cracking and from natural gas. Unconventional feedstocks like biomass and coal have not yet had a major impact on supply to the plastics industry.

The flow chart in illustrates the versatility of crude petroleum and natural gas feed stocks as starting points for the important thermosetting and thermoplastics materials. Following the first processes of crude oil distillation, naphtha feedstock is either cracked or reformed to provide useful intermediates. Thus the ethylene produced by the cracking process is of immediate use for the manufacture of polyethylene or for utilization in another process which provides a monomer, vinyl chloride—the basis of PVC. Propylene, which also arises during the cracking process, is used via either the cumene route or the isopropyl alcohol route for the manufacture of acetone needed for polymethylmethacrylate; it is also used in the manufacture of propylene oxide for polyester and polyether resins and again may be polymerized directly to polypropylene. Butenes find use in the manufacture of plasticisers and 1,3-butadiene is utilized directly for synthetic rubber manufacture. Aromatic hydrocarbons such as benzene, toluene and xylene are now widely produced from the derivatives of oil distillation operations, instead of being obtained from coal-coking processes; as the flow chart shows, these are intermediates in the manufacture of important plastics materials and auxiliary products such as plasticizers. The aromatic hydrocarbons are also a starting point for many polymers required in the synthetic fibers industry.

Production of Raw Materials Into Plastics

Many widely differing processes contribute to the final production of a finished article made wholly or partly of plastics. Some processes are purely chemical, some involve purely mechanical mixing procedures while others-particularly those towards the lower end of the diagram-involve extensive use of specialized machinery. Some of this machinery resembles that used in rubber, glass, paper and textile industries; the remainder is specific to the plastics industry.

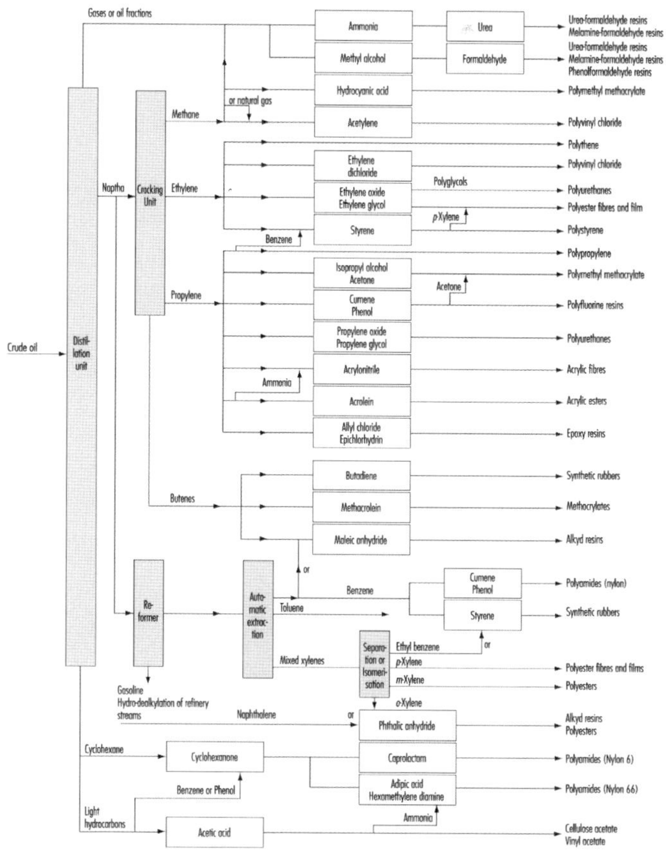

Plastics Processing

The plastics processing industry converts bulk polymeric material into finished articles.

Raw Materials

- The processing section of the plastics industry receives its raw materials for production in the following forms:
- Fully compounded polymeric material, in the form of pellets, granules or powder, which is fed directly into the machinery for processing
- Uncompounded polymer, in the form of granules or powder, which must be compounded with additives before it is suitable for feeding into to machinery
- Polymeric sheet, rod, tube and foil materials which are processed further by the industry

- Miscellaneous materials which can be fully polymerized matter in the form of suspensions or emulsions (generally known as latices) or liquids or solids which can polymerize, or substances in an intermediate state between the reactive raw materials and the final polymer. Some of these are liquids and some true solutions of partially polymerized matter in water of controlled acidity (pH) or in organic solvents.

Compounding

The manufacture of compound from polymer entails the mixing of the polymer with additives. Though a great variety of machinery is employed for this purpose, where powders are dealt with, ball mills or high-speed propeller mixers are most common, and where plastic masses are being mixed, kneading machines such as the open rolls or Banbury-type mixers, or extruders themselves are normally employed.

The additives required by the industry are many in number, and range widely in chemical type. Of some 20 classes, the most important are:

- plasticisers—generally esters of low volatility
- antioxidants—organic chemicals to protect against thermal decomposition during processing
- stabilisers—inorganic and organic chemicals to protect against thermal decomposition and against degradation from radiant energy
- lubricants
- fillers—inexpensive matter to confer special properties or to cheapen compositions
- colourants—inorganic or organic matter to color compounds
- blowing agents—gases or chemicals that emit gases to produce plastic foams.

Conversion Processes

All the conversion processes call on the "plastic" phenomenon of polymeric materials and fall into two types. Firstly, those where the polymer is brought by heat to a plastic state in which it is given a mechanical constriction leading to a form which it retains on consolidation and cooling. Secondly, those in which a polymerizable material-which may be partially polymerized is fully polymerized by the action of heat, or of a catalyst or by both acting together whilst under a mechanical constraint leading to a form which it retains when fully polymerized and cold. Plastics technology has developed to exploit these properties to produce goods with the minimum of human effort and the greatest consistency in physical properties. The following processes are commonly used.

Compression Moulding

This consists of heating a plastic material, which can be in the form of granules or powder, in a mould which is held in a press. When the material becomes "plastic" the pressure forces it to conform to the shape of the mould. If the plastic is of the type that hardens on heating, the formed article is removed after a short heating period by opening the press. If the plastic does not harden on heating, cooling must be effected before the press can be opened. Articles made by compression moulding include bottle caps, jar closures, electric plugs and sockets, toilet seats, trays and fancy goods. Compression moulding is also employed to make sheet for subsequent forming in the vacuum forming process or for building into tanks and large containers by welding or by lining existing metal tanks.

Transfer Moulding

This is a modification of compression moulding. The thermosetting material is heated in a cavity and then forced by a plunger into the mould, which is physically separate and independently heated from the heating cavity. It is preferred to normal compression moulding when the final article has to carry delicate metallic inserts such as in small electrical switchgear, or when, as in very thick objects, completion of the chemical reaction could not be obtained by normal compression moulding.

Injection Moulding

In this process, plastics granules or powders are heated in a cylinder (known as the barrel), which is separate from the mould. The material is heated until it becomes fluid, while it is conveyed through the barrel by a helical screw and then forced into the mould where it cools and hardens. The mould is then opened mechanically and the formed articles are removed. This process is one of the most important in the plastics industry. It has been extensively developed and has become capable of making articles of considerable complexity at very low cost.

Figure: An operator removing a polypropylene bowl from an injection-moulding machine.

Though transfer and injection moulding are identical in principle, the machinery employed is very different. Transfer moulding is normally restricted to thermosetting materials and injection moulding to thermoplastics.

Extrusion

This is the process in which a machine softens a plastic and forces it through a die which gives it the shape that it retains on cooling. The products of extrusion are tubes or rods which may have cross sections of almost any configuration. Tubes for industrial or domestic purposes are produced in this way, but other articles can be made by subsidiary processes. For example, sachets can be made by cutting tubes and sealing both ends, and bags from thin-walled flexible tubes by cutting and sealing one end.

The process of extrusion has two major types. In one, a flat sheet is produced. This sheet can be converted into useful goods by other processes, such as vacuum forming.

Plastic extrusion: The ribbon is chopped to make pellets for injection moulding ma-chines.

The second is a process in which the extruded tube is formed and when still hot is greatly expanded by a pressure of air maintained inside the tube. This results in a tube which can be several feet in diameter with a very thin wall. On slitting, this tube gives film which is extensively used in the packaging industry for wrapping. Alternatively the tube can be folded flat to give a two-layer sheet which can be used to make simple bags by cutting and sealing. Figure below provides an example of appropriate local ventilation on an extrusion process.

Plastic extrusion with local exhaust hood and water bath at extruder head.

Calendering

In this process, a plastic is fed to two or more heated rollers and forced into a sheet by passing through a nip between two such rollers and cooling thereafter. Sheet thicker than film is made in this way. Sheet so made is employed in industrial and domestic applications and as the raw material in the manufacture of clothing and inflated goods such as toys.

Canopy hoods to capture hot emissions from warm-up mills on a calender process

Blow Moulding

This process can be regarded as a combination of the process of extrusion and thermo-forming. A tube is extruded downwards into an opened mould; as it reaches the bottom the mould is closed round it and the tube expanded by air pressure. Thus the plastic is forced to the sides of the mould and the top and bottom sealed. On cooling, the article is taken from the mould. This process makes hollow articles of which bottles are the most important.

The compression and impact strength of certain plastic products made by blow moulding can be considerably improved by using stretch-blow moulding techniques. This is achieved by producing a pre-form which is subsequently expanded by air pressure and stretched biaxially. This has led to such an improvement in the burst pressure strength of PVC bottles that they are used for carbonated drinks.

Rotational Moulding

This process is used for the production of moulded articles by heating and cooling a hollow form which is rotated to enable gravity to distribute finely divided powder or liquid over the inner surface of that form. Articles produced by this method include footballs, dolls and other similar articles.

Film Casting

Apart from the extrusion process, films can be formed by extruding a hot polymer on to a highly polished metal drum, or a solution of polymer can be sprayed on to a moving belt.

An important application of certain plastics is the coating of paper. In this, a film of molten plastic is extruded on to paper under conditions in which the plastic adheres to the paper. Board can be coated in the same way. Paper and board so coated are widely used in packaging, and board of this type is used in box making.

Thermo-forming

Under this heading are grouped a number of processes in which a sheet of a plastic material, more often than not thermoplastic, is heated, generally in an oven, and after clamping at the perimeter is forced to a predesigned shape by pressure which may be from mechanically operated rams or by compressed air or steam. For very large articles the "rubbery" hot sheet is manhandled with tongs over formers. Products so made include external light fittings, advertising and directional road signs, baths and other toilet goods and contact lenses.

Vacuum-forming

There are many processes which come under this general heading, all of which are aspects of thermal forming, but they all have in common that a sheet of plastic is heated in a machine above a cavity, around the edge of which it is clamped, and when pliable it is forced by suction into the cavity, where it takes some specific form and cools. In a subsequent operation, the article is trimmed free from the sheet. These processes produce very cheaply thin-walled containers of all types, as well as display and advertising goods, trays and similar articles, and shock-absorbing materials for packing goods such as fancy cakes, soft fruit and cut meat.

Laminating

In all of the various laminating processes, two or more materials in the form of sheets are compressed to give a consolidated sheet or panel of special properties. At one extreme are found decorative laminates made from phenolic and amino resins, at the other complex films used in packaging having, for example, cellulose, polyethylene and metal foil in their constitution.

Resin Technology Processes

These include plywood manufacture, furniture manufacture and the construction of large and elaborate articles such as car bodies and boat hulls from glass fiber impregnated with polyester or epoxy resins. In all these processes, a liquid resin is caused to consolidate under the action of heat or of a catalyst and so bind together discrete particles or fibers or mechanically weak films or sheets, resulting in a robust panel of rigid construction. These resins can be applied by hand layup techniques such as brushing and dipping or by spraying.

Small objects such as souvenirs and plastic jewelry can also be made by casting, where the liquid resin and catalyst are mixed together and poured into a mould.

Finishing Processes

Included under this heading are a number of processes common to many industries, for example the use of paints and adhesives. There are, however, a number of specific techniques used for the welding of plastics. These include the use of solvents such as chlorinated hydrocarbons, methyl ethyl ketone (MEK) and toluene, which are used for bonding together rigid plastic sheets for general fabrication, advertising display stands and similar work. Radiofrequency (RF) radiation utilizes a combination of mechanical pressure and electromagnetic radiation with frequencies generally in the range of 10 to 100 mHz. This method is commonly used for welding together flexible plastic material in the manufacture of wallets, briefcases and children's push chairs. Ultrasonic energies are also used in combination with mechanical pressure for a similar range of work.

Types of Plastics

PET (Polyethylene Terephthalate)

PET is one of the most commonly used plastics in consumer products, and is found in most water and pop bottles, and some packaging. It is intended for single use applications; repeated use increases the risk of leaching and bacterial growth. PET plastic is difficult to decontaminate, and proper cleaning requires harmful chemicals. Polyethylene terephthalates may leach carcinogens.

PET plastic is recyclable and about 25% of PET bottles in the US today are recycled. The plastic is crushed and then shredded into small flakes which are then reprocessed to make new PET bottles, or spun into polyester fiber. This recycled fiber is used to make textiles such as fleece garments, carpets, stuffing for pillows and life jackets, and similar products.

Products made of #1 (PET) plastic should be recycled but not reused.

HDPE (High-Density Polyethylene)

HDPE plastic is the stiff plastic used to make milk jugs, detergent and oil bottles, toys, and some plastic bags. HDPE is the most commonly recycled plastic and is considered one of the safest forms of plastic. It is a relatively simple and cost-effective process to recycle HDPE plastic for secondary use.

HDPE plastic is very hard-wearing and does not break down under exposure to sunlight or extremes of heating or freezing. For this reason, HDPE is used to make picnic tables, plastic lumber, waste bins, park benches, bed liners for trucks and other products which require durability and weather-resistance.

Products made of HDPE are reusable and recyclable.

PVC (Polyvinyl Chloride)

PVC is a soft, flexible plastic used to make clear plastic food wrapping, cooking oil bottles, teething rings, children's and pets' toys, and blister packaging for myriad consumer products. It is commonly used as the sheathing material for computer cables, and to make plastic pipes and parts for plumbing. Because PVC is relatively impervious to sunlight and weather, it is used to make window frames, garden hoses, arbors, raised beds and trellises.

PVC is dubbed the "poison plastic" because it contains numerous toxins which it can leach throughout its entire life cycle. Almost all products using PVC require virgin material for their construction; less than 1% of PVC material is recycled.

Products made using PVC plastic are not recyclable. While some PCV products can be repurposed, PVC products should not be reused for applications with food or for children's use.

LDPE (Low-Density Polyethylene)

LDPE is often found in shrink wraps, dry cleaner garment bags, squeezable bottles, and the type of plastic bags used to package bread. The plastic grocery bags used in most stores today are made using LDPE plastic. Some clothing and furniture also uses this type of plastic.

LDPE is considered less toxic than other plastics, and relatively safe for use. It is not commonly recycled, however, although this is changing in many communities today as more plastic recycling programs gear up to handle this material. When recycled, LDPE plastic is used for plastic lumber, landscaping boards, garbage can liners and floor tiles. Products made using recycled LDPE are not as hard or rigid as those made using recycled HDPE plastic.

Products made using LDPE plastic are reusable, but not always recyclable. You need to check with your local collection service to see if they are accepting LDPE plastic items for recycling.

PP (Polypropylene)

Polypropylene plastic is tough and lightweight, and has excellent heat-resistance qualities. It serves as a barrier against moisture, grease and chemicals. When you try to open the thin plastic liner in a cereal box, it is polypropylene. This keeps your cereal dry and fresh. PP is also commonly used for disposable diapers, pails, plastic bottle tops, margarine and yogurt containers, potato chip bags, straws, packing tape and rope.

Polypropylene is recyclable through some curbside recycling programs, but only about 3% of PP products are currently being recycled in the US. Recycled PP is used to make landscaping border stripping, battery cases, brooms, bins and trays. However, #5 plastic is today becoming more accepted by recyclers.

PP is considered safe for reuse. To recycle products made from PP, check with your local curbside program to see if they are now accepting this material.

PS (Polystyrene)

Polystyrene is an inexpensive, lightweight and easily-formed plastic with a wide variety of uses. It is most often used to make disposable Styrofoam drinking cups, take-out "clamshell" food containers, egg cartons, plastic picnic cutlery, foam packaging and those ubiquitous "peanut" foam chips used to fill shipping boxes to protect the contents. Polystyrene is also widely used to make rigid foam insulation and underlay sheeting for laminate flooring used in home construction.

Because polystyrene is structurally weak and ultra-lightweight, it breaks up easily and is dispersed readily throughout the natural environment. Beaches all over the world have bits of polystyrene lapping at the shores, and an untold number of marine species have ingested this plastic with immeasurable consequences to their health.

Polystyrene may leach styrene, a possible human carcinogen, into food products (especially when heated in a microwave). Chemicals present in polystyrene have been linked with human health and reproductive system dysfunction.

Recycling is not widely available for polystyrene products. Most curbside collection services will not accept polystyrene, which is why this material accounts for about 35%

of US landfill material. While the technology for recycling polystyrene is available, the market for recycling is small. Awareness among consumers has grown, however, and polystyrene is being reused more often. While it is difficult to find a recycler for PS, some businesses like Mailboxes Etc. which provide shipping services are happy to receive foam packing chips for reuse.

Polystyrene should be avoided where possible.

Other (BPA, Polycarbonate and LEXAN)

The #7 category was designed as a catch-all for polycarbonate (PC) and "other" plastics, so reuse and recycling protocols are not standardized within this category. Of primary concern with #7 plastics, however, is the potential for chemical leaching into food or drink products packaged in polycarbonate containers made using BPA (Bisphenol A). BPA is a xenoestrogen, a known endocrine disruptor.

Number 7 plastics are used to make baby bottles, Sippy cups, water cooler bottles and car parts. BPA is found in polycarbonate plastic food containers often marked on the bottom with the letters "PC" by the recycling label #7. Some polycarbonate water bottles are marketed as 'non-leaching' for minimizing plastic taste or odor, however there is still a possibility that trace amounts of BPA will migrate from these containers, particularly if used to heat liquids.

A new generation of compostable plastics, made from bio-based polymers like corn starch, is being developed to replace polycarbonates. These are also included in category #7, which can be confusing to the consumer. These compostable plastics have the initials "PLA" on the bottom near the recycling symbol. Some may also say "Compostable."

#7 plastics are not for reuse, unless they have the PLA compostable coding. When possible it is best to avoid #7 plastics, especially for children's food. Plastics with the recycling labels #1, #2 and #4 on the bottom are safer choices and do not contain BPA. PLA coded plastics should be thrown in the compost and not the recycle bin since PLA compostable plastics are not recyclable.

The plastics industry has conformed to regulations by applying the required codes to consumer products, but it is up to individuals to read and understand the codes.

Environmental Impacts of Plastic

1. Plastic is ubiquitous

Plastic debris is found absolutely everywhere, from the Arctic to Antarctica. It clogs street drains in our cities; it litters campgrounds and national parks, and is even piling up on Mount Everest (link). But thanks runoff, and to our fondness for directly dumping our trash into the nearest river or lake, plastic is growing increasingly common in the world's oceans.

A crab on the beach of Henderson Island uses a piece of plastic debris as a home. When researchers traveled to the tiny, uninhabited island in the middle of the Pacific Ocean, they were astonished to find an estimated 38 million pieces of trash washed up on the beaches.

Mountains of plastic trash have been found everywhere in the world's oceans, from one of the remotest specks of dirt on the planet, Henderson Island, a tiny uninhabited coral atoll in the middle of the Pacific Ocean , to the deepest spot on Earth, the Mariana Trench , which plummets to a depth of 10,994 meters (36,070 feet) below sea level. As that flotilla of escaped rubber duckies demonstrated, floating plastic even forms massive "garbage patches" swirling slowly in the middle of the Pacific Ocean, thousands of kilometers away from land.

But more than floating around in the water column, plastic trash is found in the guts of more than 90% of the world's sea birds , in the stomachs of more than half of the world's sea turtles , and it's even choking the life out of whales. At the rate at which plastic is accumulating in the oceans of the planet, it's predicted that, by 2050, the mass of plastic in the world's oceans will exceed the mass of all the fish that live there.

2. Plastics are one of the main products of fracking

We already know that fracking is bad for the planet -- it pollutes water, soil and air with toxins, it creates underground cavities that collapse into sinkholes, and it raises pressure in underground rock formations, destabilizing them and leading to earthquakes, even in places where earthquakes are uncommon. Adding insult to injury, one of the main products of fracking is plastics. Basically, fossil fuels removed from shale and other rock formations are turned into resin pellets that are used to manufacture ever more plastics, plastics that are easily discarded, plastics that usually are designed to be single use.

You may not know that plastics manufacturers are dangerous to the environment, to wildlife and to people in other ways, too. For example, the British chemical industry giant, Ineos, has 75 manufacturing facilities in 22 countries, including the United States. In addition to contributing to growing mountains of plastic rubbish, Ineos facilities can boast an impressive -- and growing -- record of fires, explosions, and chemical leaks.

"The Ineos facility in Cologne, Germany was the site of a massive fire in 2008, and there have been a series of high profile accidents elsewhere a major oil leak in Norway, a number of chemical leaks in France, long-running controversies over chemical dumping in Italy, and the release of toxic gas that resulted in the hospitalization of workers in Belgium," according to a report written by Food and Water Watch and Food and Water Europe.

3. Plastics kill more than just people

In the last few months, the effects upon wildlife that come from eating, or becoming entangled in, plastic debris have been reported more widely and more often than ever before, leading to public outcry and protests.

The unaltered stomach contents of a dead albatross chick on Midway Atoll National Wildlife Refuge in the Pacific in September 2009 include plastic marine debris fed to the chick by its parents.

These tragic events should come as no surprise: there are an estimated 270,000 tons of plastic floating through the world's seas where it threatens 700 marine species with its presence. Further, there is growing evidence that plastics play a role in rising rates of species extinctions.

But entangling or lodging inside the digestive tract of an unfortunate victim, like whales and other marine mammals, birds, sea turtles and fishes, is not the end of the line, because plastics do not biodegrade, nor are they digestible. Instead, large plastic pieces break into smaller fragments that are easy for even more animal species to consume. Eating plastics results in malnutrition, intestinal blockage, or slow poisoning from chemicals leached from or attached to plastics.

It's almost as if seabirds and other marine animals are intentionally seeking out plastics and consuming them -- but why? Recent research proposed that plastics in the oceans become covered with marine algae that release a natural sulfur compound, dimethylsulfoniopropionate (DMSP), when they die or become damaged. This chemical is detected by hungry seabirds and other marine creatures, which are attracted to it and then they mistake plastics for lunch.

Not even microscopic animals, collectively known as zooplankton, are safe from ingesting plastics, the only difference being that they consume microscopic plastic bits. Similar to plastics consumption by larger animals, microplastics can result in reduced feeding, energetic deficiencies, injury, or death of zooplankton -- A Very Bad Thing since zooplankton are part of the essential foundation upon which the entire marine food web rests.

4. Not all plastic is recyclable and not all recyclable plastic is recycled

People are often confused by the terms, "break down" versus "biodegradable" (or "compostable"). When plastics are broken down, this simply means one large piece of plastic is reduced into a bunch of smaller pieces of plastic. These smaller pieces of plastic can be consumed by smaller animals, but are still indigestible.

A minority of plastics are "compostable" or "biodegradable", which means they can be reduced to their chemical components by, say, your home compost. Other plastics can only be successfully composted by industrial or municipal facilities after first being separated from other, non-biodegradable, plastics.

Unfortunately, most people are confused as to which plastics fall into this biodegradable category, and this confusion often is most apparent when confronted with unfamiliar plastics in unfamiliar places. For example, are plastic water bottles purchased, say, at the airport biodegradable? Should they be disposed into the recycling bin or into the food waste bin? It's often challenging to know, especially since little useful information is typically available to quickly make a decision.

5. Most plastics last forever

Once again, as demonstrated by the 1992 rubber duckie escape, plastics survive even the harshest conditions, such as floating around in a marine environment under blistering, unrelenting sunshine or frozen into Arctic ice for years before finally floating away and landing on some faraway shore. For this reason, plastics will probably outlast humanity itself. It is this apparent immortality leads me to my last point, which is that plastics may be co-opted for good.

Construction Industry

Construction industry is the branch of manufacturing and trade based on the building, maintaining, and repairing structures. This includes drilling and solid mineral exploration. It is a booming industry and remain so with the continuation of the development process especially in the developing countries. With the process of development, the migration of people takes place from the rural to urban areas. This phenomenon is most significantly observed in the "Asian Tiger" countries, China and India. Thus, the Construction Industry is also on a rise in such countries.

Construction Industry Categories:

Building Construction Industry: All general contractors and operative builders primarily engaged in the construction of residential, farm, industrial, commercial, or other buildings.

Heavy Construction Industry: All general contractors primarily engaged in heavy construction other than building, such as highways and streets, bridges, sewers, railroads, irrigation projects, and flood control projects and marine construction. This includes special trade contractors primarily engaged in activities not normally performed on buildings, such as highway grading or underwater rock removal. This does not include special trade contractors primarily engaged activities performed on buildings.

Special Trade Construction Industry: All special trade contractors who undertake activities of a type that are specialized either to building construction, including work on mobile homes, or to both building and nonbuilding projects. This includes projects such as painting, electrical work, plumbing, etc. This does not include activities specialized for heavy construction.

New Construction Techniques And Sustainability

The development of efficiency codes has prompted the development of new construction technologies and methods, many pioneered by academic departments of construction management that seek to improve efficiency and performance while reducing construction waste.

New techniques of building construction are being researched, made possible by advances in 3D printing technology. In a form of additive building construction, similar to the additive manufacturing techniques for manufactured parts, building printing is making it possible to flexibly construct small commercial buildings and private habitations in around 20 hours, with built-in plumbing and electrical facilities, in one continuous build, using large 3D printers. Working versions of 3D-printing building technology are already printing 2 meters (6 ft 7 in) of building material per hour as of January 2013, with the next-generation printers capable of 3.5 meters (11 ft) per hour, sufficient to complete a building in a week.

In the current trend of sustainable construction, the recent movements of New Urbanism and New Classical Architecture promote a sustainable approach towards construction, that appreciates and develops smart growth, architectural tradition and classical design. This is in contrast to modernist and short-lived globally uniform architecture, as well as opposing solitary housing estates and suburban sprawl. Both trends started in the 1980s.

Wood is being introduced as a realistic material for skyscrapers (nicknamed "plyscrapers") thanks to new developments in super-strong engineered timber, whose collective name is "mass timber" and includes cross-laminated timber.

Processes

Shasta Dam under construction in June 1942

Design Team

In the industrialized world, construction usually involves the translation of designs into reality. A formal design team may be assembled to plan the physical proceedings, and to integrate those proceedings with the other parts. The design usually consists of drawings and specifications, usually prepared by a design team including Architect, civil engineers, mechanical engineers, electrical engineers, structural engineers, fire protection engineers, planning consultants, architectural consultants, and archaeological consultants. The design team is most commonly employed by (i.e. in contract with) the property owner. Under this system, once the design is completed by the design team, a number of construction companies or construction management companies may then be asked to make a bid for the work, either based directly on the design, or on the basis of drawings and a bill of quantities provided by a quantity surveyor. Following evaluation of bids, the owner typically awards a contract to the most cost efficient bidder.

The best modern trend in design is toward integration of previously separated specialties, especially among large firms. In the past, architects, interior designers, engineers, developers, construction managers, and general contractors were more likely to be entirely separate companies, even in the larger firms. Presently, a firm that is nominally an "architecture" or "construction management" firm may have experts from all related fields as employees, or to have an associated company that provides each necessary skill. Thus, each such firm may offer itself as "one-stop shopping" for a construction project, from beginning to end. This is designated as a "design build" contract where the contractor is given a performance specification and must undertake the project from design to construction, while adhering to the performance specifications.

Several project structures can assist the owner in this integration, including design-build, partnering and construction management. In general, each of these project structures allows the owner to integrate the services of architects, interior designers, engineers and constructors throughout design and construction. In response, many companies are growing beyond traditional offerings of design or construction services alone and are placing more emphasis on establishing relationships with other necessary participants through the design-build process.

The increasing complexity of construction projects creates the need for design professionals trained in all phases of the project's life-cycle and develop an appreciation of the building as an advanced technological system requiring close integration of many sub-systems and their individual components, including sustainability. Building engineering is an emerging discipline that attempts to meet this new challenge.

Financial Advisors

Trump International Hotel and Tower (Chicago)

May 23, 2006

September 14, 2007 (3 months before completion)

Construction projects can suffer from preventable financial problems. Underbids happen when builders ask for too little money to complete the project. Cash flow problems exist when the present amount of funding cannot cover the current costs for labour and materials, and because they are a matter of having sufficient funds at a specific time, can arise even when the overall total is enough. Fraud is a problem in many fields, but is notoriously prevalent in the construction field. Financial planning for the project is intended to ensure that a solid plan with adequate safeguards and contingency plans are in place before the project is started and is required to ensure that the plan is properly executed over the life of the project.

Mortgage bankers, accountants, and cost engineers are likely participants in creating an overall plan for the financial management of the building construction project. The presence of the mortgage banker is highly likely, even in relatively small projects since the owner›s equity in the property is the most obvious source of funding for a building project. Accountants act to study the expected monetary flow over the life of the project and to monitor the payouts throughout the process. Cost engineers and estimators apply expertise to relate the work and materials involved to a proper valuation. Cost overruns with government projects have occurred when the contractor identified change orders or project changes that increased costs, which are not subject to competition from other firms as they have already been eliminated from consideration after the initial bid.

Large projects can involve highly complex financial plans and often start with a conceptual estimate performed by a building estimator. As portions of a project are completed, they may be sold, supplanting one lender or owner for another, while the logistical requirements of having the right trades and materials available for each stage of the building construction project carries forward. In many English-speaking countries, but not the United States, projects typically use quantity surveyors.

Construction along Ontario Highway 401, widening the road from six to twelve travel lanes

A construction project must fit into the legal framework governing the property. These include governmental regulations on the use of property, and obligations that are created in the process of construction.

When applicable, the project must adhere to zoning and building code requirements. Constructing a project that fails to adhere to codes does not benefit the owner. Some legal requirements come from malum in se considerations, or the desire to prevent indisputably bad phenomena, e.g. explosions or bridge collapses. Other legal requirements come from malum prohibitumconsiderations, or factors that are a matter of custom or expectation, such as isolating businesses from a business district or residences from a residential district. An attorney may seek changes or exemptions in the law that governs the land where the building will be built, either by arguing that a rule is inapplicable (the bridge design will not cause a collapse), or that the custom is no longer needed (acceptance of live-work spaces has grown in the community).

A construction project is a complex net of contracts and other legal obligations, each of which all parties must carefully consider. A contract is the exchange of a set of obligations between two or more parties, but it is not so simple a matter as trying to get the other side to agree to as much as possible in exchange for as little as possible. The time element in construction means that a delay costs money, and in cases of bottlenecks, the delay can be extremely expensive. Thus, the contracts must be designed to ensure that each side is capable of performing the obligations set out. Contracts that set out clear expectations and clear paths to accomplishing those expectations are far more likely to result in the project flowing smoothly, whereas poorly drafted contracts lead to confusion and collapse.

Legal advisors in the beginning of a construction project seek to identify ambiguities and other potential sources of trouble in the contract structure, and to present options for preventing problems. Throughout the process of the project, they work to avoid and resolve conflicts that arise. In each case, the lawyer facilitates an exchange of obligations that matches the reality of the project.

Interaction of Expertise

Apartment complex under construction in Daegu, South Korea

Design, finance, and legal aspects overlap and interrelate. The design must be not only structurally sound and appropriate for the use and location, but must also be financially possible to build, and legal to use. The financial structure must accommodate the need for building the design provided, and must pay amounts that are legally owed. The legal structure must integrate the design into the surrounding legal framework, and enforce the financial consequences of the construction process.

Procurement

Procurement describes the merging of activities undertaken by the client to obtain a building. There are many different methods of construction procurement; however the three most common types of procurement are traditional (design-bid-build), design-build and management contracting.

There is also a growing number of new forms of procurement that involve relationship contracting where the emphasis is on a co-operative relationship among the principal, the contractor, and other stakeholders within a construction project. New forms include partnering such as Public-Private Partnering (PPPs) aka private finance initiatives (PFIs) and alliances such as “pure” or “project” alliances and “impure” or “strategic” alliances. The focus on co-operation is to ameliorate the many problems that arise from the often highly competitive and adversarial practices within the construction industry.

Traditional

This is the most common method of construction procurement and is well established and recognized. In this arrangement, the architect or engineer acts as the project co-ordinator. His or her role is to design the works, prepare the specifications and produce construction drawings, administer the contract, tender the works, and manage the works from inception to completion. There are direct contractual links between the architect›s client and the main contractor. Any subcontractor has a direct contractual relationship with the main contractor. The procedure continues until the building is ready to occupy.

Design-build

This approach has become more common in recent years, and also involves the client contracting a single entity that both provides a design and builds it. In some cases, the design-build package can also include finding the site, arranging funding and applying for all necessary statutory consents.

Construction of the Phase-1 (first two towers) of the Havelock City Project, Sri Lanka

The owner produces a list of requirements for a project, giving an overall view of the project's goals. Several D&B contractors present different ideas about how to accomplish these goals. The owner selects the ideas they like best and hires the appropriate contractor. Often, it is not just one contractor, but a consortium of several contractors working together. Once these have been hired, they begin building the first phase of the project. As they build phase 1, they design phase 2. This is in contrast to a design-bid-build contract, where the project is completely designed by the owner, then bid on, then completed.

Kent Hansen pointed out that state departments of transportation usually use design build contracts as a way of progressing projects when states lack the skills-resources. In such departments, design build contracts are usually employed for very large projects.

Management Procurement Systems

In this arrangement the client plays an active role in the procurement system by entering into separate contracts with the designer (architect or engineer), the construction manager, and individual trade contractors. The client takes on the contractual role, while the construction or project manager provides the active role of managing the separate trade contracts, and ensuring that they complete all work smoothly and effectively together.

Management procurement systems are often used to speed up the procurement processes, allow the client greater flexibility in design variation throughout the contract, give the ability to appoint individual work contractors, separate contractual responsibility on each individual throughout the contract, and to provide greater client control.

In recent time, construction software starts to get traction – as it digitizes construction industry. Among solutions, there are for example: Procore, GenieBelt, PlanGrid, bouw7, etc.

Authority Having Jurisdiction

Digging the foundation for a building construction in Jakarta, Indonesia

In construction, the authority having jurisdiction (AHJ) is the governmental agency or sub-agency that regulates the construction process. In most cases, this is the municipality where the building is located. However, construction performed for supra-municipal authorities are usually regulated directly by the owning authority, which becomes the AHJ.

Construction on the Federal Reserve building in Kansas City, Missouri

Before the foundation can be dug, contractors are typically required to verify and have existing utility lines marked, either by the utilities themselves or through a company specializing in such services. This lessens the likelihood of damage to the existing electrical, water, sewage, phone, and cable facilities, which could cause outages and potentially hazardous situations. During the construction of a building, the municipal building inspector inspects the building periodically to ensure that the construction adheres to the approved plans and the local building code. Once construction is complete and a final inspection has been passed, an occupancy permit may be issued.

An operating building must remain in compliance with the fire code. The fire code is enforced by the local fire department or a municipal code enforcement office.

Changes made to a building that affect safety, including its use, expansion, structural integrity, and fire protection items, usually require approval of the AHJ for review concerning the building code.

Safety

At-risk workers without appropriate safety equipment

Construction is one of the most dangerous occupations in the world, incurring more occupational fatalities than any other sector in both the United States and in the European Union. In 2009, the fatal occupational injury rate among construction workers in the United States was nearly three times that for all workers, with Falls being one of the most common causes of fatal and non-fatal injuries among construction workers. Proper safety equipment such as harnesses, hard hats and guardrails and procedures such as securing ladders and inspecting scaffolding can curtail the risk of occupational injuries in the construction industry. Other major causes of fatalities in the construction industry include electrocution, transportation accidents, and trench cave-ins.

Roofing requires a very high level of safety

Other safety risks for workers in construction include hearing loss due to high noise exposure, musculoskeletal injury, chemical exposure, and high levels of stress. Construction has been identified by the National Institute for Occupational Safety and Health (NIOSH) as a priority industry sector in the National Occupational Research Agenda (NORA) to identify and provide intervention strategies regarding occupational health and safety issues.

Construction projects around the world have a significant impact on our environment, both on a local and a global scale. Every stage of the construction process has a measurable environmental impact: the mining processes used to source materials, the transportation of these materials to the building site from sources around the world, the construction process itself and the waste removal and disposal process that follows the completion of the project. With a rapidly developing global economy, it is crucial that we understand how the construction projects we undertake impact the environment and how we can measure and reduce that impact in the future.

Construction's Impact on the Environment and Climate Change

Climate change scientists believe that the Earth is rapidly getting warmer and that human activity is playing a significant role in accelerating this process. The biggest way in which construction firms are doing this is by contributing to carbon dioxide emissions. Carbon dioxide is a gaseous product of burning fossil fuels like gas and diesel. This gas gets trapped in our atmosphere, creating a "greenhouse effect" that warms the Earth over a long period of time.

Given that, every construction project results in the emission of carbon dioxide, methane, and other waste products that pollute the air and are believed to contribute to global climate change. The most harmful aspect of construction in terms of contributing to climate change is the operation of heavy machinery in mining projects that extract raw materials from the Earth for use in construction projects. The global cement industry contributes approximately 5% of global carbon dioxide emissions. Applications of fuel and electricity are also major contributors – fossil fuels are used to extract and transport minerals, to process materials, and even to power tools on construction sites.

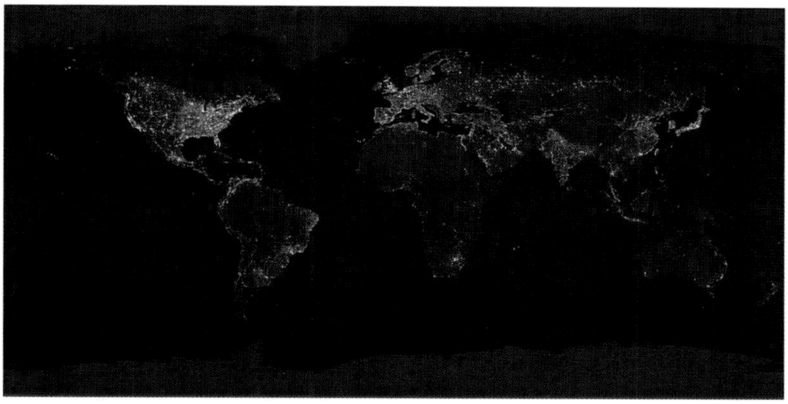

Environmental Impacts of Construction Projects

Aside from contributing to climate change on a global scale, individual construction projects can have a significant impact on local environments and nature. There are numerous sources of water pollution on building sites, including diesel and other fossil fuels, paints, solvents, and toxic chemicals. Even minor chemical spills leech into the ground and may enter waterways where they toxify water and harm aquatic life. In the developing world, there are often less stringent requirements for the disposal of toxic construction waste products, resulting in significant local environmental damage following a building project.

Even greater than the impacts of the construction process itself is the environmental impact of the actual buildings that are being constructed. It has been estimated that the daily operation of buildings accounts for 40% of total energy usage worldwide. Projects like the Dakota Access Pipeline have been criticized for their long-term environmental impact on the areas they run through. Critics of this project suspect that transporting fossil fuels through hundreds of miles of pipeline will lead to oil spills that can go undetected for long periods and damage or destroy wild lands that were previously undisturbed.

Oil Industry

Oil Industry, also known as the Petroleum Industry, includes exploration, production, refining, transportation of crude and petroleum Products via rail, road, pipelines or sea (oil tankers) and marketing of petroleum products. It plays very vital role in the global economy of any nation because it promotes industrialization.

The Oil Industry is broadly classified into three sectors:

- Upstream
- Midstream
- Downstream

The upstream sector includes well completions and drilling operations. Exploration and production organizations perform 2D and 3D seismic surveys of the ground to search for hydrocarbon formations of crude oil and natural gas. Once such formations are discovered or explored, exploratory wells are drilled and subsequently drilling is carried out. Some of the upstream sector organizations are: Chevron, BP, Shell, Exxon-Mobil, ConocoPhillips, Total and many more.

The midstream sector involves the transportation of oil or refined petroleum products to various refineries or industries for further processing or usage mainly through rail, tankers and pipelines.

The Downstream sector includes refineries that process crude oil into petroleum products such as LPG, Naphtha, Motor Spirit, High Speed Diesel, Jet Kerosene, Furnace Oil, and heavy products. Once these products are produced, refineries marketing agencies or Oil Marketing Companies (OMC's) sell these petroleum products to various geographical locations.

Industry Structure

The modern petroleum industry is structured into four types of organizations:

1. state enterprises,

2. multinational corporations,
3. the Organization of Petroleum Exporting Countries (OPEC), and
4. legally organized commodity exchanges. In most petroleum-producing countries in the early twenty-first century, the state owns a petroleum company with a legal monopoly on the production and distribution of crude petroleum. The largest of these after 1918 was the Soviet state petroleum enterprise.

Most crude petroleum is produced by Western companies. In 1870 the Standard Oil Company was created by the Rockefeller brothers and two other partners. In 1893 the Standard Oil Trust was formed in New Jersey to evade an antitrust suit brought by the state of Ohio. Gulf Oil Corporation was formed in 1901 by the Richard K. Mellon family. Texaco was founded as the Texas Company in 1901 by Joseph S. Cullinan, Walter B. Sharp, and Arnols Shaect. The U.S. Supreme Court in 1911 ordered Standard Oil Trust dissolved. The component companies, still with the plurality of shares owned by the former Rockefeller shareholders, continued to operate as ostensibly separate companies. These companies were Standard Oil of Ohio, Standard Oil of Indiana, Standard Oil of New York, Standard Oil of New Jersey, Standard Oil of California, Standard Oil of Kentucky, Atlantic, and the Ohio Oil Company (later Marathon).

The Royal Dutch Company for the Exploration of Petroleum Sources in the Netherlands Indies was established in 1890. It merged in 1907 with Shell Transport and Trading Company, a British company, to form Royal Dutch Shell. The British government formed the Anglo-Iranian Oil Company in 1914. This company ultimately became British Petroleum. In 1924 the Compagnie Française des Pétroles was established by the French government. ELF began before World War II (1939–1945) with the establishment of three small companies to explore for gas near oil seepages in Aquitaine. Italy in 1926 formed Agencia Generale Italiani Petroli (AGIP). In 1953 ENI was founded as a conglomerate of thirty-six subsidiaries, including AGIP.

These companies all were multinational, multidivisional, and vertically and horizontally integrated firms. Within each company, five major functions were performed—exploration, drilling and production, transportation, refining, and distribution to final consumers. Subsidiaries were responsible for each function. Transfers of product among these subsidiaries and other divisions of the companies were accomplished using shadow pricing or some other form of transfer pricing. Actual payments were made only for transactions with outside firms.

OPEC was established in Baghdad in September 1960 as an intergovernmental organization of five original member states—Iraq, Iran, Kuwait, Saudi Arabia, and Venezuela. Its charter required that each member acquire an increasing level of control of production. By 1970 each member state was required to own a minimum of 55 percent of foreign petroleum companies operating within its jurisdiction. Iraqi production has not been part of OPEC quota agreements since March 1998 due to U.S. and United

Nations controls. Other OPEC members have included Qatar (joined 1961), Indonesia and Libya(1962), Ecuador (1963–1993), Trucial States of Oman (now United Arab Emirates, 1967), Algeria(1969), Nigeria (1971), Gabon (1975–1995), and Angola (2007). The OPEC cartel was formed to control the world oil supply so as to increase revenue to member states. It operates by assigning members an annual supply quota for crude oil production and export. In 2005 it controlled about 41.7 percent of world production. OPEC also sets prices.

Environmental Impacts of Oil Industry

Toxicity

Petroleum distillates can create a sheen on the surface of water as a thin layer creating an optical phenomenon called interphase.

Petroleum is a complex mixture of many components. These components include straight chained , branched , cyclic , monocyclic aromatic and polycyclic aromatic hydrocarbons. The toxicity of oils can be understood using the toxic potential or the toxicity of each individual component of oil at the water solubility of that component. There are many methods that can be used to measure the toxicity of crude oil and other petroleum related products. Certain studies analyzing levels of toxicity can use the target lipid model or colorimetric analysis using colored-dyes in order to assess toxicity and biodegradability.

Different oils and petroleum related products have different levels of toxicity. Levels of toxicity are influenced by many factors such as weathering , solubility , as well as chemical properties such as persistence. Increased weathering tends to decrease levels of toxicity as more soluble and lower molecular weight substances are removed. Highly soluble substances tend to have higher levels of toxicity than substances that are not very soluble in water. Generally oils that have longer carbon chains and with more benzene rings have higher levels of toxicity. Benzene is the petroleum related product with the highest level of toxicity. Other substances other than benzene which are highly toxic are toluene, methylbenzene and xylenes (BETX).Substances with the lowest toxicity are crude oil and motor oil.

Despite varying levels of toxicity amongst different variants of oil , all petroleum -derived products have adverse impacts on human health and the ecosystem. Examples of adverse effects are oil emulsions in digestive systems in certain mammals might result in decreased ability to digest nutrients that might lead to death of certain mammals. Further symptoms include capillary ruptures and hemorrhages. Ecosystem food chains can be affected due to a decrease in algae productivity therefore threatening certain species. Oil is "acutely lethal" to fish - that is, it kills fish quickly, at a concentration of 4000 parts per million (ppm) (0.4%). The toxicity of petroleum related products threaten human health. Many compounds found in oil are highly toxic and can cause cancer (carcinogenic) as well as other diseases. Studies in Taiwan link proximity to oil refineries to premature births. Crude oil and petroleum distillates cause birth defects.

Benzene is present in both crude oil and gasoline and is known to cause leukaemia in humans. The compound is also known to lower the white blood cell count in humans, which would leave people exposed to it more susceptible to infections. "Studies have linked benzene exposure in the mere parts per billion (ppb) range to terminal leukemia, Hodgkin›s lymphoma, and other blood and immune system diseases within 5-15 years of exposure."

Exhaust

Petroleum diesel exhaust from a truck

Emissions from the petroleum industry occur in every chain of the oil producing process from the extraction to the consumption phase. In the extraction phase there are emissions of not only carbon dioxide but various other pollutants like nitrous oxidesand aerosols. Certain by - products include carbon monoxide and methanol. When oil or petroleum distillates are combusted , usually the combustion is not complete and the chemical reaction leaves by-products which are not water or carbon dioxide. However despite the large amounts of pollutants there is uncertainty about the amount and concentration of certain pollutants. In the refinement stages of petroleum also contributes to large amounts of pollution in urban areas. This increase of pollution has adverse effects on human health due to the toxicity of oil. A study investigating the effects of oil refineries in Taiwan. The study found an increased occurrence of pre-mature births in mothers that lived in close proximity to oil refineries than mothers who lived away from oil refineries. There were also differences observed in sex ratios and the birth weight of the children. Also, fine particulates of soot blacken humans› and other animals› lungs and cause heart problems or death. Soot is cancer causing (carcinogenic).

Acid Rain

Trees killed by acid rain, an unwanted side effect of burning petroleum

The combustion process of petroleum , coal , and wood is responsible for increased occurrence of acid rain. Combustion causes an increased amount of nitrous oxides , along with sulfur dioxide from the sulfur in the oil. These by products combine with water in the atmosphere to create acid rain. The increased concentrations of nitrates and other acidic substances have significant effects on the pH levels of rainfall. Data samples analyzed from the United States and Europe from the past 100 years and showed an increase in nitrous oxide emissions from combustion. The emissions were large enough to acidify the rainfall. The acid rain has adverse impacts on the larger ecosystem. For example acid rain can kill trees , acidified lakes and dead fish. Coral reefs are also destroyed from the acidified rainfall from acid rain. Acid rain also leads to the corrosion of machinery and structures (large amounts of capital) and to the slow destruction of archeological structures like the marble ruins of Rome and Greece.

Climate Change

The combustion of petroleum causes an increased amount of carbon dioxide emissions as well as other greenhouse gases. The first study on the effects of carbon dioxide was studied by Swedish Nobel chemist Svante Arrhenius. His mathematical model showed that an increase of carbon dioxide results in an increase in surface temperatures, therefore both factors are correlated. The combustion of petroleum for transport, industrial and domestic use is one of the major forms of air pollution. The ultimate by-product of oil combustion is carbon dioxide, however there are other by-products such as monoxide and nitrates. These by-products react with the atmosphere to produce ozone and other greenhouse gases. The increased pollution has consequences on global temperature. The atmosphere reflects 30% of the incoming long wave radiation back and keeps 70% of it for warmth. However, an increased carbon dioxide concentration in the atmosphere acts as a "blanket" to increased heat. Therefore, more long wave radiation is trapped in the atmosphere when there is a higher concentration of carbon dioxide and this trapping results in increased surface temperatures. IPCC (2007) states that the climate system will heat up by 3 degrees Celsius for a doubling of carbon dioxide. The warming of the temperatures will have massive impacts on rainfall patterns , retreat of glaciers , and the average sea levels.

Oil Spills

A Bird Covered In Oil From The Black Sea Oil Spill.

An oil spill is the release of a liquid petroleum hydrocarbon into the environment, especially marine areas, due to human activity, and is a form of pollution. The term is usually applied to marine oil spills, where oil is released into the ocean or coastal waters, but spills may also occur on land. Oil spills may be due to releases of crude oil from tankers, pipelines, railcars, offshore platforms, drilling rigs and wells, as well as spills of refined petroleum products (such as gasoline, diesel) and their by-products, heavier fuels used by large ships such as bunker fuel, or the spill of any oily refuse or waste oil.

Major oil spills include , Lakeview Gusher, Gulf War oil spill, and the Deep-water Horizon oil spill. Spilt oil penetrates into the structure of the plumage of birds and the fur of mammals, reducing its insulating ability, and making them more vulnerable to temperature fluctuations and much less buoyant in the water. Cleanup and recovery from an oil spill is difficult and depends upon many factors, including the type of oil spilled, the temperature of the water (affecting evaporation and biodegradation), and the types of shorelines and beaches involved. Other factors influencing the rate of long-term contamination is the continuous inputs of petroleum residues and the rate at which the environment can clean itself Spills may take weeks, months or even years to clean up.

Volatile Organic Compounds

Volatile organic compounds (VOCs) are gases or vapours emitted by various solids and liquids." Petroleum hydrocarbons such as gasoline, diesel, or jet fuel intruding into indoor spaces from underground storage tanks or brownfields threaten safety (e.g., explosive potential) and causes adverse health effects from inhalation.

Waste Oil

Waste oil in the form of motor oil

Waste oil is oil containing not only breakdown products but also impurities from use. Some examples of waste oil are used oils such as hydraulic oil, transmission oil, brake fluids, motor oil, crankcase oil, gear box oil and synthetic oil. Many of the same problems associated with natural petroleum exist with waste oil. When waste oil from vehicles drips out engines over streets and roads, the oil travels into the water table bringing with it such toxins as benzene. This poisons both soil and drinking water. Runoff from storms carries waste oil into rivers and oceans, poisoning them as well.

Agricultural Industry

Agriculture, known also as husbandry or farming, is the science of cultivating plants, animals, and other life forms for food, fiber, and fuel. The agricultural industry, which includes enterprises engaged in growing crops, raising fish and animals, and logging wood, encompasses farms, dairies, hatcheries, and ranches. It is considered to be a major industry.

The development of agriculture spans thousands of years and has been affected by human cultures, climate variations, and evolving technologies. Despite the industry's evolution over the centuries, all types of farming still rely on methods to maintain environments conducive for raising domestic species. In terms of plants, the land must be sustained with a type of irrigation system; for raising livestock, rangeland must be cultivated to support animal life. Nowadays farm yields have greatly increased due to improvements in breeding, chemicals, and technology. Yet some of these measures can cause damage to the environment and pose health risks to humans.

A modern innovation in the agriculture industry is the Green Revolution, which began in the second half of the 20th century. This development is cited for saving people around the world from starvation through initiatives involving irrigation systems, production of high-yielding crops, and better management strategies.

Today farming professionals face the daunting challenge of keeping current with increasingly rapid changes in equipment technology and agricultural research. For instance, biotechnology is being applied to industrial livestock operations to provide healthier animals and improve breed development, resulting in more meat, eggs, and dairy products to meet consumer demands.

Responsible for putting healthy food on consumers' tables, the agriculture industry is regulated by strict policies. Farmers must be up to date on all legislation affecting their product and their laborers. For instance, the 2012 Agriculture Reform, Food and Jobs Act provided the most significant reforms in farming policy in years. It stopped direct subsidy payments, added new dairy and livestock programs, and strengthened crop insurance.

A farmer may work a single tract of land in addition to renting out a number of separate tracts. The farmer may be the sole operator of the establishment or work within a partnership or corporation. The agriculture industry employs a wide-range of workers including farm laborers, veterinarians, scientists, salespeople, and soil managers.

Despite all the advances made in the agriculture field, it still remains a hazardous industry. Fatalities and injuries due to machine operation as well as diseases and birth defects spawned by exposure to pesticides and fertilizers are among the work-related dangers faced by agricultural workers.

A hard-hitting recession and severe droughts have plagued the agriculture industry over the past decade. While employment in some sectors of this industry is expected to rise, other areas are forecasted to decline. For instance, opportunities are promising for scientists and research and development technicians but disappointing for farmers, ranchers, and laborers.

Early Agriculture

Early people depended for their survival on hunting, fishing, and food gathering. To this day, some groups still pursue this simple way of life, and others have continued as roving herders. However, as various groups of people undertook deliberate cultivation of wild plants and domestication of wild animals, agriculture came into being. Cultivation of crops notably grains such as wheat, rice, corn, rye, barley, and millet encouraged settlement of stable farm communities, some of which grew to be towns and city-states in various parts of the world. Early agricultural implements the digging stick, the hoe, the scythe, and the plow developed slowly over the centuries, each innovation (e.g., the introduction of iron) causing profound changes in human life. From early times, too, people created ingenious systems of irrigation to control water supply, especially in semiarid areas and regions of periodic rainfall, e.g., the Middle East, the American Southwest and Mexico, the Nile Valley, and S Asia.

Farming was often intimately associated with landholding and therefore with political organization. Growth of large estates involved the use of slaves and bound or semi free labor. In the Western Middle Ages the manorial system was the typical organization of more or less isolated units and determined the nature of the agricultural village. In Asia large holdings by the nobles, partly arising from feudalism (especially in China and Japan), produced a similar pattern.

The Rise of Commercial Agriculture

As the Middle Ages waned, increasing communications, the commercial revolution, and the rise of cities in Western Europe tended to turn agriculture away from subsistence farming toward the growing of crops for sale outside the community (commercial agriculture). In Britain the practice of in closure allowed landlords to set aside plots of land, formerly subject to common rights, for intensive cropping or fenced pasturage, leading to efficient production of single crops.

In the 16th and 17th century horticulture was greatly developed and contributed to the so-called agricultural revolution. Exploration and intercontinental trade, as well as scientific investigation, led to the development of horticultural knowledge of various crops and the exchange of farming methods and products, such as the potato, which was introduced from America along with beans and corn (maize) and became almost as common in N Europe as rice is in SE Asia.

The appearance of mechanical devices such as the sugar mill and Eli Whitney's cotton gin helped to support the system of large plantations based on a single crop. The Industrial Revolution after the late 18th century swelled the population of towns and cities and increasingly forced agriculture into greater integration with general economic and financial patterns. In the American colonies the independent, more or less self-sufficient family farm became the norm in the North, while the plantation, using slave labor, was dominant (although not universal) in the South. The free farm pushed westward with the frontier.

Modern Agriculture

In the N and W United States the era of mechanized agriculture began with the invention of such farm machines as the reaper, the cultivator, the thresher, and the combine. Other revolutionary innovations, e.g., the tractor continued to appear over the years, leading to a new type of large-scale agriculture. Modern science has also revolutionized food processing; refrigeration, for example, has made possible the large meatpacking plants and shipment and packaging of perishable foods. Urbanization has fostered the specialties of market gardening and truck farming. Harvesting operations have been mechanized for almost every plant product grown. Breeding programs have developed highly specialized animal, plant, and poultry varieties, thus increasing production efficiency. The development of genetic engineering has given rise to genetically modified transgenic crops and, to a lesser degree, livestock that possess a gene from an unrelated species that confers a desired quality. Such modification allows livestock to be used as "factories" for the production of growth hormone and other substances In the

United States and other leading food-producing nations agricultural colleges and government agencies attempt to increase output by disseminating knowledge of improved agricultural practices, by the release of new plant and animal types, and by continuous intensive research into basic and applied scientific principles relating to agricultural production and economics.

These changes have, of course, given new aspects to agricultural policies. In the United States and other developed nations, the family farm is disappearing, as industrialized farms, which are organized according to industrial management techniques, can more efficiently and economically adapt to new and ever-improving technology, specialization of crops, and the volatility of farm prices in a global economy. Niche farming, in which specialized crops are raised for a specialized market, e.g., heirloom tomatoes or exotic herbs sold to gourmet food shops and restaurants, revived or encouraged some smaller farms in the latter 20th and early 21st cents., but did little to stop the overall decrease in family farms. In Third World countries, where small farms, using rudimentary techniques, still predominate, the international market has had less effect on the internal economy and the supply of food.

Most of the governments of the world face their own type of farm problem, and the attempted solutions vary as much as does agriculture itself. The modern world includes areas where specialization and conservation have been highly refined, such as Denmark, as well as areas such as N Brazil and parts of Africa, where forest peoples employ "slash-and-burn" agriculture—cutting down and burning trees, exhausting the ash-enriched soil, and then moving to a new area. In other regions, notably SE Asia, dense population and very small holdings necessitate intensive cultivation, using people and animals but few machines; here the yield is low in relation to energy expenditure. In many countries extensive government programs control the planning, financing, and regulation of agriculture. Agriculture is still the occupation of almost 50% of the world's population, but the numbers vary from less than 3% in industrialized countries to over 60% in Third World countries.

Industrialized Agriculture

Industrialized agriculture is the type of agriculture where large quantities of crops and livestock are produced through industrialized techniques for the purpose of sale. The goal of industrialized agriculture is to increase crop yield, which is the amount of food that is produced for each unit of land. Crops and livestock made through this type of agriculture are produced to feed the masses and the products are sold worldwide.

Industrialized agriculture is able to produce large quantities of food due to the farming methods used. Instead of using animal and manpower to work the fields, industrialized agriculture utilizes large machines, which are more powerful and can work faster and harder. The shift towards machines has increased the use of fossil fuels on industrial farms, and, therefore, the price of food can fluctuate as the price of oil changes. Industrialized agriculture also increases crop yield by investing in large irrigation systems and by using chemical fertilizers and pesticides.

The chemical fertilizers that are used in industrialized agriculture often add inorganic nutrients to the soil to increase yield and plant size. The use of pesticides is also common in industrialized agriculture, and most pesticides help increase yield by killing pests that are harming or consuming the crops. Another farming technique that is used in industrialized agriculture is the method of growing monocultures, which is when a single crop is planted on a large scale. Although planting monocultures can increase overall yield, this method of farming is also more susceptible to disease and causes a reduction in the dietary variation of consumers.

Subsistence Agriculture

Although industrialized agriculture is necessary to feed the growing human population, there is another type of agriculture that is regularly practiced today. Subsistence agriculture is when a farmer lives on a small amount of land and produces enough food to feed his or her household and have a small cash crop. The goal of subsistence agriculture is to produce enough food to ensure the survival of the individual family. If there is excess food produced, it is sold locally to other families or individuals.

Subsistence agriculture varies a great deal from industrialized agriculture in terms of the farming methods used. This type of agriculture is very labor-intensive because all of the work is done by humans and animals and only hand tools and simple machines are used to work the land.

Subsistence agriculture does not rely on chemical fertilizers or pesticides and instead utilizes more natural techniques. Most farmers have animals, including chickens, goats and cows, and the manure from these animals is used to fertilize the plants. The crops produced are then consumed or sold, and the inedible parts of the plants are used to feed the livestock. This creates a closed circuit within the farm where nothing goes to waste.

Instead of using chemical pesticides, subsistence farmers rely on natural predators of pests to control the pest population. Another major difference between industrialized and subsistence agriculture is what is being planted. Unlike industrialized agriculture that utilizes monocultures, subsistence agriculture relies on polycultures, which is when different types of crops are planted in one area. Planting polycultures is a method used to get the most crop yield out of a small area of land.

Environmental Impacts of Agriculture

Climate Change

Agriculture and climate change have a reciprocal relationship. Climate change affects agricultural production through precipitation levels and temperature variations. In return, poor agricultural practices increase climate change. The most significant climate change associated with agriculture is brought about by methane, nitrous oxide and carbon dioxide, all of which are greenhouse gases released into the Earth's atmosphere from farming. Pesticides and fertilizers application affect the quality of air through production of compounds like phosphorus, nitrate and ammonia.

Pollution

In order to enhance agricultural production quantity and quality, several additives to the soil are used in farming. The widely used are pesticides and fertilizers, which end up as pollutants in water run-off from the soil. This run-off can adversely affect more people and animal wildlife.

Soil Degradation

In all ecosystems, the biodiversity held in soil is massive. Healthy soils are vital to creating ample food production. Although agriculture is not the sole cause of soil degradation, poor farming practices are known to cause a considerable decline in in the quality of soil. This mainly results from pesticide contamination, waterlogging and salting. Soil erosion leads to loss of soil fertility and structure.

Deforestation

Across the globe, cases of forests being cleared for agricultural reasons is on the rise with people seeking to increase their scales of agricultural production. Deforestation provides more land for crops and pasture. Deforestation enhances the effects of climate change. Destruction of habitat amongst species also leads to fragmentation and depletion. Extensive deforestation affects the water cycle, which results in interferences with precipitation.

Ways To Minimize Environmental Impacts Of Agriculture

The key to protecting the environment against the harmful effects of extensive and unsafe agricultural practices is through sustainable practices. Sustainable farming incorporates both the conservation of available resources as well as employing farming practices aimed at environmental protection.

References

- Hamilton, Michael S. (2005). Mining environmental policy: Comparing Indonesia and the USA. Ashgate studies in environmental policy and practice. Burlington, VT: Ashgate Publishing. ISBN 978-0-7546-4493-4
- Mining-operation-types-impacts-and-remedial-measures-30038: yourarticlelibrary.com, Retrieved 26 March 2018
- Gabbard, Alex (5 February 2008). "Coal Combustion: Nuclear Resource or Danger". Oak Ridge National Laboratory. Archived from the original on 5 February 2007. Retrieved 2008-10-22
- What-is-the-environmental-impact-of-the-mining-industry: worldatlas.com, Retrieved 16 July 2018

- Ek, A. S.; Renberg, I. (2001). "Heavy metal pollution and lake acidity changes caused by one thousand years of copper mining at Falun, central Sweden". Journal of paleolimnology. 26 (1): 89–107
- Economic-significance-of-mining: avalaresources.com, Retrieved 29 June 2018
- Shrader-Frechette, Kristin. What Will Work: Fighting Climate Change with Renewable Energy, Not Nuclear Power, Oxford University Press, 2011, pg.9, ISBN 0-19-979463-4
- Environmental-impacts-of-construction-projects: esub.com, Retrieved 09 May 2018
- Kharecha P.A.; Hansen J.E. (2008). "Implications of "peak oil" for atmospheric CO_2 and climate». Global Biogeochem. Cycles. 22: GB3012. arXiv:0704.2782. Bibcode:2008GBioC..22.3012K. doi:10.1029/2007GB003142
- Oil-industry-7092: petropedia.com, Retrieved 30 March 2018
- "Coal Fatalities for 1900 Through 2016". Arlington, VA: U.S. Mine Safety and Health Administration (MSHA). Retrieved 2017-10-25
- Types-of-agriculture-industrialized-and-subsistence-agriculture: study.com, Retrieved 09 May 2018
- RYAN, PADDY A. (1991). "Environmental effects of sediment on New Zealand streams: a review". New Zealand Journal of Marine and Freshwater Research. 25: 207–221. doi:10.1080/00288330.1991.9516472
- What-is-the-environmental-impact-of-agriculture: worldatlas.com, Retrieved 14 April 2018

Occupational Health Hazards and Safety

Any hazard experienced in the workplace is called an occupational hazard. The topics elaborated in this chapter cover the different aspects related to occupation health hazards and safety, such as occupational stress, work accident, occupational injury, occupational noise, occupational heat stress and occupational dust exposure, besides many others.

Occupational health and safety (OHS) relates to health, safety, and welfare issues in the workplace. OHS includes the laws, standards, and programs that are aimed at making the workplace better for workers, along with co-workers, family members, customers, and other stakeholders. Improving a company's occupational health and safety standards ensures good business, a better brand image, and higher employee morale. Generally, occupational health and safety aims at the promotion and maintenance of the highest degree of physical, mental and social well-being of workers in all occupations;

- the prevention among workers of adverse effects on health caused by their working conditions;
- the protection of workers in their employment from risks resulting from factors adverse to health;
- the placing and maintenance of workers in an occupational environment adapted to physical and mental needs;
- the adaptation of work to humans.

In other words, occupational health and safety encompasses the social, mental and physical well-being of workers that is the "whole person".

Successful occupational health and safety practice requires the collaboration and participation of both employers and workers in health and safety programmes, and involves the consideration of issues relating to occupational medicine, industrial hygiene, toxicology, education, engineering safety, ergonomics, psychology, etc.

Occupational health issues are often given less attention than occupational safety issues because the former are generally more difficult to confront. However, when health is addressed, so is safety, because a healthy workplace is by definition also a safe workplace. The converse, though, may not be true - a so-called safe workplace is not necessarily also a healthy workplace. The important point is that issues of both health and safety must be addressed in every workplace. By and large, the definition of occupational health and safety given above encompasses both health and safety in their broadest contexts

Workplace Hazards

Although work provides many economic and other benefits, a wide array of workplace hazards also present risks to the health and safety of people at work. These include but are not limited to, "chemicals, biological agents, physical factors, adverse ergonomic conditions, allergens, a complex network of safety risks," and a broad range of psychosocial risk factors. Personal protective equipment can help protect against many of these hazards.

Physical hazards affect many people in the workplace. Occupational hearing loss is the most common work-related injury in the United States, with 22 million workers exposed to hazardous noise levels at work and an estimated $242 million spent annually on worker's compensation for hearing loss disability. Falls are also a common cause of occupational injuries and fatalities, especially in construction, extraction, transportation, healthcare, and building cleaning and maintenance. Machines have moving parts, sharp edges, hot surfaces and other hazards with the potential to crush, burn, cut, shear, stab or otherwise strike or wound workers if used unsafely.

Biological hazards (biohazards) include infectious microorganisms such as viruses and toxins produced by those organisms such as anthrax. Biohazards affect workers in many industries; influenza, for example, affects a broad population of workers. Outdoor workers, including farmers, landscapers, and construction workers, risk exposure to numerous biohazards, including animal bites and stings, urushiol from poisonous plants, and diseases transmitted through animals such as the West Nile virus and Lyme disease. Health care workers, including veterinary health workers, risk exposure to blood-borne pathogens and various infectious diseases, especially those that are emerging.

Dangerous chemicals can pose a chemical hazard in the workplace. There are many classifications of hazardous chemicals, including neurotoxins, immune agents, dermatologic agents, carcinogens, reproductive toxins, systemic toxins, asthmagens, pneumoconiotic agents, and sensitizers. Authorities such as regulatory agencies set occupational exposure limits to mitigate the risk of chemical hazards. An international effort is investigating the health effects of mixtures of chemicals. There is some evidence that certain chemicals are harmful at lower levels when mixed with one or more other chemicals. This may be particularly important in causing cancer.

Psychosocial hazards include risks to the mental and emotional well-being of workers, such as feelings of job insecurity, long work hours, and poor work-life balance. A recent Cochrane review - using moderate quality evidence - related that the addition of work-directed interventions for depressed workers receiving clinical interventions reduces the number of lost work days as compared to clinical interventions alone. This review also demonstrated that the addition of cognitive behavioral therapy to primary or occupational care and the addition of a "structured telephone outreach and care management program" to usual care are both effective at reducing sick leave days.

By Industry

Specific occupational safety and health risk factors vary depending on the specific sector and industry. Construction workers might be particularly at risk of falls, for instance, whereas fishermen might be particularly at risk of drowning. The United States Bureau of Labor Statistics identifies the fishing, aviation, lumber, metalworking, agriculture, mining and transportation industries as among some of the more dangerous for workers. Similarly psychosocial risks such as workplace violence are more pronounced for certain occupational groups such as health care employees, police, correctional officers and teachers.

Construction

Construction is one of the most dangerous occupations in the world, incurring more occupational fatalities than any other sector in both the United States and in the European Union. In 2009, the fatal occupational injury rate among construction workers in the United States was nearly three times that for all workers. Falls are one of the most common causes of fatal and non-fatal injuries among construction workers. Proper safety equipment such as harnesses and guardrails and procedures such as securing ladders and inspecting scaffolding can curtail the risk of occupational injuries in the construction industry. Due to the fact that accidents may have disastrous consequences for employees as well as organizations, it is of utmost importance to ensure health and safety of workers and compliance with HSE construction requirements. Health and safety legislation in the construction industry involves many rules and regulations. For example, the role of the Construction Design Management (CDM) Coordinator as a requirement has been aimed at improving health and safety on-site.

Workplace safety notices at the entrance of a Chinese construction site.

The 2010 National Health Interview Survey Occupational Health Supplement (NHIS-OHS) identified work organization factors and occupational psychosocial and chemical/physical exposures which may increase some health risks. Among all U.S. workers in the construction sector, 44% had non-standard work arrangements (were not regular permanent employees) compared to 19% of all U.S. workers, 15% had temporary employment compared to 7% of all U.S. workers, and 55% experienced job insecurity compared to 32% of all U.S. workers. Prevalence rates for exposure to physical/chemical hazards were especially high for the construction sector. Among nonsmoking workers, 24% of construction workers were exposed to secondhand smoke while only 10% of all U.S. workers were exposed. Other physical/chemical hazards with high prevalence rates in the construction industry were frequently working outdoors (73%) and frequent exposure to vapors, gas, dust, or fumes (51%).

Agriculture

Agriculture workers are often at risk of work-related injuries, lung disease, noise-induced hearing loss, skin disease, as well as certain cancers related to chemical use or prolonged sun exposure. On industrialized farms, injuries frequently involve the use of agricultural machinery. The most common cause of fatal agricultural injuries in the United States is tractor rollovers, which can be prevented by the use of roll over protection structures which limit the risk of injury in case a tractor rolls over. Pesticides and other chemicals used in farming can also be hazardous to worker health, and workers exposed to pesticides may experience illnesses or birth defects. As an industry in which families, including children, commonly work alongside their families, agriculture is a common source of occupational injuries and illnesses among younger workers. Common causes of fatal injuries among young farm worker include drowning, machinery and motor vehicle-related accidents.

The 2010 NHIS-OHS found elevated prevalence rates of several occupational exposures in the agriculture, forestry, and fishing sector which may negatively impact health.

These workers often worked long hours. The prevalence rate of working more than 48 hours a week among workers employed in these industries was 37% and 24% worked more than 60 hours a week. Of all workers in these industries, 85% frequently worked outdoors compared to 25% of all U.S. workers. Additionally, 53% were frequently exposed to vapors, gas, dust, or fumes, compared to 25% of all U.S. workers.

Rollover protection bar on a Fordson tractor.

Service Sector

As the number of service sector jobs has risen in developed countries, more and more jobs have become sedentary, presenting a different array of health problems than those associated with manufacturing and the primary sector. Contemporary problems such as the growing rate of obesity and issues relating to occupational stress, workplace bullying, and overwork in many countries have further complicated the interaction between work and health.

According to data from the 2010 NHIS-OHS, hazardous physical/chemical exposures in the service sector were lower than national averages. On the other hand, potentially harmful work organization characteristics and psychosocial workplace exposures were relatively common in this sector. Among all workers in the service industry, 30% experienced job insecurity in 2010, 27% worked non-standard shifts (not a regular day shift), 21% had non-standard work arrangements (were not regular permanent employees).

Due to the manual labour involved and on a per employee basis, the US Postal Service, UPS and FedEx are the 4th, 5th and 7th most dangerous companies to work for in the US.

Mining and Oil & Gas Extraction

According to data from the 2010 NHIS-OHS, workers employed in mining and oil and gas extraction industries had high prevalence rates of exposure to potentially harmful work organization characteristics and hazardous chemicals. Many of these workers worked long hours: 50% worked more than 48 hours a week and 25% worked more than 60 hours a week in 2010. Additionally, 42% worked non-standard shifts (not a regular day shift). These workers also had high prevalence of exposure to physical/ chemical hazards. In 2010, 39% had frequent skin contact with chemicals. Among non-smoking workers, 28% of those in mining and oil and gas extraction industries had frequent exposure to secondhand smoke at work. About two-thirds were frequently exposed to vapors, gas, dust, or fumes at work.

Healthcare and Social Assistance

Beekeepers often wear protective clothing, for OHS reasons

Healthcare workers are exposed to many hazards that can adversely affect their health and well-being. Long hours, changing shifts, physically demanding tasks, violence, and exposures to infectious diseases and harmful chemicals are examples of hazards that put these workers at risk for illness and injury.

According to the Bureau of Labor statistics, U.S. hospitals recorded 253,700 work-related injuries and illnesses in 2011, which is 6.8 work-related injuries and illnesses for every 100 full-time employees. The injury and illness rate in hospitals is higher than the rates in construction and manufacturing – two industries that are traditionally thought to be relatively hazardous.

The Occupational Health Safety Network (OHSN) is a secure electronic surveillance system developed by the National Institute for Occupational Safety and Health (NIOSH) to address health and safety risks among health care personnel. OHSN uses existing data to characterize risk of injury and illness among health care workers. Hospitals and other healthcare facilities can upload the occupational injury data they already collect to the secure database for analysis and benchmarking with other de-identified facilities. NIOSH works with OHSN participants in identifying and implementing timely and targeted interventions. OHSN modules currently focus on three high risk and preventable events that can lead to injuries or musculoskeletal disorders among healthcare personnel: musculoskeletal injuries from patient handling activities; slips, trips, and falls; and workplace violence. OHSN enrollment is open to all healthcare facilities.

Workplace Fatality And Injury Statistics

United States

Parts of this article (those related to No info here from post-2015) need to be **updated**. Please update this article to reflect recent events or newly available information. *(March 2018)*

The Bureau of Labor Statistics of the United States Department of Labor compiles information about workplace fatalities and non-fatal injuries in the United States. In 1970, an estimated 14,000 workers were killed on the job – by 2010, the workforce had doubled, but workplace deaths were down to about 4,500. Between 1913 and 2013, workplace fatalities dropped by approximately 80%.

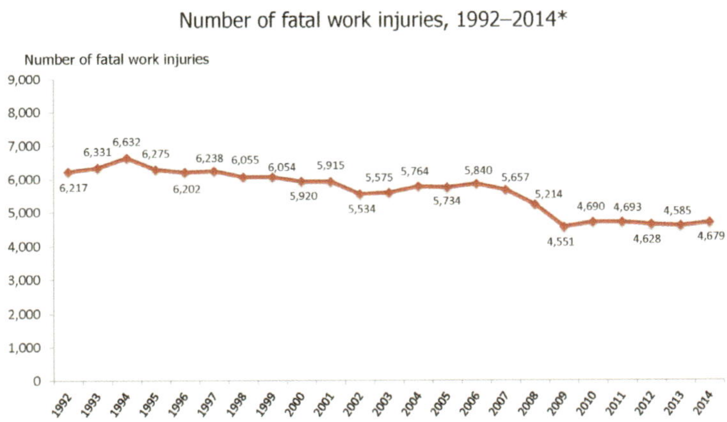

Number of occupational fatal work injuries in the U.S. from 1992 until 2014. Note, 2001 statistics do not include death related to the September 11 terrorist attacks.

The Bureau also compiles information about the most dangerous jobs. According to the census of occupational injuries 4,679 people died on the job in 2014. In 2015, a decline in nonfatal workplace injuries and illnesses was observed, with private industry employers reporting approximately 2.9 million incidents, nearly 48,000 fewer cases than in 2014. The Bureau also uses online tools to identify and compile additional sources of fatality reports for their datasets.

Job	Fatalities	Fatalities per 100,000 employees
Fishermen	22	80.8
Pilots	81	63.2
Timber cutter	77	109.5
Structural metal workers	15	25.2
Waste collectors	27	35.8
Farmers and ranchers	263	26.0
Power-line workers	25	19.2
Construction workers and miners	130	17.9
Roofers	81	46.2
Truck drivers and other drivers	835	23.4
All occupations	4,679	3.3

2014 employer-reported injuries and illnesses		
Industry	**Rate per 100 full-time employees**	**Number**
Agriculture, forestry, fishing and hunting	5.5	52,400
Mining, quarrying, and oil and gas extraction	2.0	17,900
Construction (private)	3.6	200,900
Manufacturing	4.0	483,300
Wholesale trade	2.9	163,100
Retail trade	3.6	416,100
Transportation and warehousing (private)	4.8	201,500
Utilities (private)	2.4	13,400
Information	1.4	35,300
Finance and insurance	0.7	34,800
Real estate, rental, and leasing	2.9	51,100
Professional, scientific, and technical services	0.9	69,900
Management of companies and enterprises	1.0	21,100
Administrative and support services	2.4	109,300
Waste management and remediation services (private)	5.1	19,900
Educational services (private)	2.1	38,500
Health care and social assistance (private)	4.5	612,500
Leisure and hospitality	3.6	337,500
State government: Education	4.1	31,100
State government: Health care and social assistance	8.1	43,800
State government: Justice, public order, and safety activities	6.1	43,200

2014 employer-reported injuries and illnesses		
Industry	**Rate per 100 full-time employees**	**Number**
Local government: Construction	8.6	8,700
Local government: Transportation and warehousing	7.5	17,900
Local government: Utilities	5.4	12,200
Local government: Education	4.1	225,100
Local government: Health care and social assistance	5.6	41,000
Local government: Justice, public order, and safety activities	9.5	84,200
All industries including state and local government	3.4	3,675,800

Cause of injury and illness	2014 rate per 10,000 full-time employees
Contact with objects	23.8
Fall to lower level	5.4
Fall on same level	18.8
Slips or trips without fall	4.4
Over-exertion in lifting/lowering	11.0
Repetitive motion	2.7
Exposure to harmful substances or environments	4.3
Transportation incidents	5.8
Fires and explosions	0.2
Violence and other injuries by person or animal	6.8
Total	107.1

Musculoskeletal injuries accounted for 32% of all employer-reported injuries and illnesses in 2014.

European Union

In most countries males comprise the vast majority of workplace fatalities. In the EU as a whole, 94% of death were of males. In the UK the disparity was even greater with males comprising 97.4% of workplace deaths. In the UK there were 171 fatal injuries at work in financial year 2011–2012, compared with 651 in calendar year 1974; the fatal injury rate declined over that period from 2.9 fatalities per 100,000 workers to 0.6 per 100,000 workers

Management Systems

National

National management system standards for occupational health and safety include AS/NZS 4801-2001 for Australia and New Zealand, CAN/CSA-Z1000-14 for Canada and ANSI/ASSE Z10-2012 for the United States. Association Française de Normalisation

(AFNOR) in France also developed occupational safety and health management standards. In the United Kingdom, non-departmental public body Health and Safety Executive published Managing for health and safety (MFHS), an online guidance. In Germany, the state factory inspectorates of Bavaria and Saxony had introduced the management system OHRIS. In the Netherlands, the management system Safety Certificate Contractors combines management of occupational health and safety and the environment.

International

Iso 45001 Was Published In March 2018.

Previously, the International Labour Organization (ILO) published ILO-OSH 2001, also titled "Guidelines a on occupational safety and health management systems" to assist organizations with introducing OSH management systems. These guidelines encourage continual improvement in employee health and safety, achieved via a constant process of policy, organization, planning & implementation, evaluation, and action for improvement, all supported by constant auditing to determine the success of OSH actions. From 1999 to 2018, the occupational health and safety management system standard OHSAS 18001 was adopted as a British and Polish standard and widely used internationally. OHSAS 18000 comprised two parts, OHSAS 18001 and 18002 and was developed by a selection of leading trade bodies, international standards and certification bodies to address a gap where no third-party certifiable international standard existed. It was intended to integrate with ISO 9001 and ISO 14001.

National Legislation And Public Organizations

Occupational safety and health practice vary among nations with different approaches to legislation, regulation, enforcement, and incentives for compliance. In the EU, for example, some member states promote OSH by providing public monies as subsidies, grants or financing, while others have created tax system incentives for OSH investments. A third group of EU member states has experimented with using workplace accident insurance premium discounts for companies or organizations with strong OSH records.

In Australia, the Commonwealth, four of the six states and both territories has enacted and administer harmonised Work Health and Safety Legislation in accordance with the Intergovernmental Agreement for Regulatory and Operational Reform in Occupational Health and Safety. Each of these jurisdictions has enacted Work Health & Safety legislation and regulations based on the Commonwealth Work Health and Safety Act 2011 and common Codes of Practice developed by Safe Work Australia. Some jurisdictions have also included mine safety under the model approach, however, most have retained separate legislation for the time being. Western Australia intends to adopt a moderated version of the model approach and Victoria has retained its own regime, although the

Model WHS laws themselves drew heavily on the Victoria approach.

In Canada, workers are covered by provincial or federal labor codes depending on the sector in which they work. Workers covered by federal legislation (including those in mining, transportation, and federal employment) are covered by the Canada Labor Code; all other workers are covered by the health and safety legislation of the province in which they work. The Canadian Centre for Occupational Health and Safety (CCOHS), an agency of the Government of Canada, was created in 1966 by an Act of Parliament. The act was based on the belief that all Canadians had "a fundamental right to a healthy and safe working environment." CCOHS is mandated to promote safe and healthy workplaces to help prevent work-related injuries and illnesses. The CCOHS maintains a useful (partial) list of OSH regulations for Canada and its provinces.

Number of full-time OSH inspectors per 100,000 full-time employees	
Italy	17.7
Finland	17.5
Denmark	11.9
United Kingdom	11.1
Norway	10.6
Sweden	10
Belgium	5.3
Netherlands	4.8
Ireland	4.5
Greece	4.1
France	3.5
Spain	2.1

In the European Union, member states have enforcing authorities to ensure that the basic legal requirements relating to occupational health and safety are met. In many EU countries, there is strong cooperation between employer and worker organizations (e.g. unions) to ensure good OSH performance as it is recognized this has benefits for both the worker (through maintenance of health) and the enterprise (through improved productivity and quality). In 1996, the European Agency for Safety and Health at Work was founded.

Member states of the European Union have all transposed into their national legislation a series of directives that establish minimum standards on occupational health and safety. These directives (of which there are about 20 on a variety of topics) follow a similar structure requiring the employer to assess the workplace risks and put in place preventive measures based on a hierarchy of control. This hierarchy starts with elimination of the hazard and ends with personal protective equipment.

However, certain EU member states admit to having lacking quality control in occupational safety services, to situations in which risk analysis takes place without any on-site workplace visits and to insufficient implementation of certain EU OSH directives. Based on this, it is hardly surprising that the total societal costs of work-related health problems and accidents vary from 2.6% to 3.8% of GNP between the EU member states.

In Denmark, occupational safety and health is regulated by the Danish Act on Working Environment and cooperation at the workplace. The Danish Working Environment Authority carries out inspections of companies, draws up more detailed rules on health and safety at work and provides information on health and safety at work. The result of each inspection is made public on the web pages of the Danish Working Environment Authority so that the general public, current and prospective employees, customers and other stakeholders can inform themselves about whether a given organization has passed the inspection, should they wish to do so.

In Spain, occupational safety and health is regulated by the Spanish Act on Prevention of Labor Risks. The Ministry of Employment and Social Security is the authority responsible for issues relating to labor environment. The National Institute for Labor Safety and Hygiene is the technical public Organization specialized in occupational safety and health.

In Sweden, occupational safety and health is regulated by the Work Environment Act. The Swedish Work Environment Authority is the government agency responsible for issues relating to the working environment. The agency should work to disseminate information and furnish advice on OSH, has a mandate to carry out inspections, and a right to issue stipulations and injunctions to any non-compliant employer.

In the UK, health and safety legislation is drawn up and enforced by the Health and Safety Executive and local authorities (the local council) under the Health and Safety at Work etc. Act 1974 (HASAWA). HASAWA introduced a general duty on an employer to ensure, so far as is reasonably practicable, the health, safety and welfare at work of all his employees; with the intention of giving a legal framework supporting 'codes of practice' not in themselves having legal force but establishing a strong presumption as to what was reasonably practicable (deviations from them could be justified by appropriate risk assessment). The previous reliance on detailed prescriptive rule-setting was seen as having failed to respond rapidly enough to technological change, leaving new technologies potentially un-regulated or inappropriately regulated. HSE has continued to make some regulations giving absolute duties (where something must be done with no 'reasonable practicability' test) but in the UK the regulatory trend is away from prescriptive rules, and towards 'goal setting' and risk assessment. Recent major changes to the laws governing asbestos and fire safety management embrace the concept of risk assessment. The other key aspect of the UK legislation is a statutory mechanism for worker involvement through elected health and safety representatives and health and safety committees. This followed a similar approach in Scandinavia, and that approach has since been adopted in Australia, Canada, New Zealand and Malaysia, for example.

For the UK, the government organization dealing with occupational health has been the Employment Medical Advisory Service but in 2014 a new occupational health organization - the Health and Work Service - was created to provide advice and assistance to employers in order to get back to work employees on long-term sick-leave. The service, funded by government, will offer medical assessments and treatment plans, on a voluntary basis, to people on long-term absence from their employer; in return, the government will no longer foot the bill for Statutory Sick Pay provided by the employer to the individual.

In India, the Labour Ministry formulates national policies on occupational safety and health in factories and docks with advice and assistance from Directorate General of Factory Advice Service and Labor Institutes (DGFASLI), and enforces its Policies through inspectorates of factories and inspectorates of dock safety. DGFASLI is the technical arm of the Ministry of Labor & Employment, Government of India and advises the factories on various problems concerning safety, health, efficiency and well - being of the persons at work places. The DGFASLI provides technical support in formulating rules, conducting occupational safety surveys and also for conducting occupational safety training programs.

In Malaysia, the Department of Occupational Safety and Health (DOSH) under the Ministry of Human Resource is responsible to ensure that the safety, health and welfare of workers in both the public and private sector is upheld. DOSH is responsible to enforce the Factories and Machinery Act 1967 and the Occupational Safety and Health Act 1994.

People's Republic of China

In the People's Republic of China, the Ministry of Health is responsible for occupational disease prevention and the State Administration of Work Safety for safety issues at work. On the provincial and municipal level, there are Health Supervisions for occupational health and local bureaus of Work Safety for safety. The "Occupational Disease Control Act of PRC" came into force on May 1, 2002. and Work safety Act of PRC on November 1, 2002. The Occupational Disease Control Act is under revision. The prevention of occupational disease is still in its initial stage compared with industrialized countries such as the US or UK.

In Singapore, the Ministry of Manpower operates various checks and campaigns against unsafe work practices, such as when working at height, operating cranes and in traffic management. Examples include Operation Cormorant and the Falls Prevention Campaign.

Hardware stores in China specializing in safety equipment

In South Africa the Department of Labour is responsible for occupational health and safety inspection and enforcement in commerce and industry apart from mining and energy production, where the Department of Mineral Resources is responsible.

The main statutory legislation on Health and Safety in the jurisdiction of the Department of Labor is Act No. 85 of 1993: Occupational Health and Safety Act as amended by Occupational Health and Safety Amendment Act, No. 181 of 1993.

Regulations to the OHS Act include:

- General Administrative Regulations, 2003
- Certificate of Competency Regulations, 1990
- Construction Regulations, 2014
- Diving Regulations 2009
- Driven Machinery Regulations, 1988
- Environmental Regulations for Workplaces, 1987
- General Machinery regulations, 1988
- General Safety Regulations, 1986
- Noise induced hearing loss regulations, 2003
- Pressure Equipment Regulations, 2004

In Taiwan, the Occupational Safety and Health Administration of the Ministry of Labor is in charge of occupational safety and health. The matter is governed under the Occupational Safety and Health Act.

2007, official release the document TOSHMS (Taiwan Occupational Safety and Health Management System), which defined the basic rule about occupational safety standard.

OSHAD is introduced on 2010 to regulate the implementation of occupational health and safety in the region.

In the United States, President Richard Nixon signed the Occupational Safety and Health Act into law on December 29, 1970. The act created the three agencies that administer it. They include the Occupational Safety and Health Administration, National Institute for Occupational Safety and Health, and the Occupational Safety and Health Review Commission. The act authorized the Occupational Safety and Health Administration (OSHA) to regulate private employers in the 50 states, the District of Columbia, and territories. The Act establishing it includes a general duty clause (29 U.S.C. § 654, 5(a)) requiring an employer to comply with the Act and regulations derived from it, and to provide employees with "employment and a place of employment which are free from recognized hazards that are causing or are likely to cause death or serious physical harm to his employees."

OSHA was established in 1971 under the Department of Labor. It has headquarters in Washington, DC and ten regional offices, further broken down into districts, each organized into three sections; compliance, training, and assistance. Its stated mission is to assure safe and healthful working conditions for working men and women by setting and enforcing standards and by providing training, outreach, education and assistance. The original plan was for OSHA to oversee 50 state plans with OSHA funding 50% of each plan. Unfortunately it has not worked out that way. There are currently 26 approved state plans (4 cover only public employees) and no other states want to participate. OSHA manages the plan in the states not participating.

OSHA develops safety standards in the Code of Federal Regulation and enforces those safety standards through compliance inspections conducted by Compliance Officers; enforcement resources are focussed on high-hazard industries. Worksites may apply to enter OSHA's Voluntary Protection Program (VPP); a successful application leads to an on-site inspection ; if this is passed the site gains VPP status and OSHA no longer inspect it annually nor (normally) visit it unless there is a fatal accident or an employee complaint until VPP revalidation (after 3–5 years)(VPP sites have injury and illness rates less than half the average for their industry).

It has 73 specialists in local offices to provide tailored information and training to employers and employees at little or no cost Similarly OSHA produces a range of publications, provides advice to employers and funds consultation services available for small businesses.

OSHA's Alliance Program enables groups committed to worker safety and health to work with it to develop compliance assistance tools and resources, share information with workers and employers, and educate them about their rights and responsibilities. OSHA also has a Strategic Partnership Program that zeros in on specific hazards or specific geographic areas. OSHA manages Susan B. Harwood grants to nonprofit companies to train workers and employers to recognize, avoid, and prevent safety and health hazards in the workplace. Grants focus on small business, hard-to-reach workers and high-hazard industries.

The National Institute for Occupational Safety and Health, created under the same act, works closely with OSHA and provides the research behind many of OSHA's regulations and standards.

Professional Roles And Responsibilities

The roles and responsibilities of OSH professionals vary regionally, but may include evaluating working environments, developing, endorsing and encouraging measures that might prevent injuries and illnesses, providing OSH information to employers, employees, and the public, providing medical examinations, and assessing the success of worker health programs.

In Norway, the main required tasks of an occupational health and safety practitioner include the following:

- Systematic evaluations of the working environment
- Endorsing preventative measures which eliminate causes of illnesses in the workplace
- Providing information on the subject of employees' health
- Providing information on occupational hygiene, ergonomics, and environmental and safety risks in the workplace

In the Netherlands, the required tasks for health and safety staff are only summarily defined and include the following:

- Providing voluntary medical examinations
- Providing a consulting room on the work environment to the workers
- Providing health assessments (if needed for the job concerned)

'The main influence of the Dutch law on the job of the safety professional is through the requirement on each employer to use the services of a certified working conditions service to advise them on health and safety'. A 'certified service' must employ sufficient numbers of four types of certified experts to cover the risks in the organisations which use the service:

- A safety professional
- An occupational hygienist
- An occupational physician
- A work and organisation specialist.

In 2004, 37% of health and safety practitioners in Norway and 14% in the Netherlands had an MSc; 44% had a BSc in Norway and 63% in the Netherlands; and 19% had training as an OSH technician in Norway and 23% in the Netherlands.

Leather craftsman gloves, safety goggles, and a properly fitted hardhat are crucial for proper safety in a construction environment.

The main tasks undertaken by the OHS practitioner in the US include:

- Develop processes, procedures, criteria, requirements, and methods to attain the best possible management of the hazards and exposures that can cause injury to people, and damage property, or the environment;
- Apply good business practices and economic principles for efficient use of resources to add to the importance of the safety processes;
- Promote other members of the company to contribute by exchanging ideas and other different approaches to make sure that every one in the corporation possess OHS knowledge and have functional roles in the development and execution of safety procedures;
- Assess services, outcomes, methods, equipment, workstations, and procedures by using qualitative and quantitative methods to recognise the hazards and measure the related risks;

- Examine all possibilities, effectiveness, reliability, and expenditure to attain the best results for the company concerned.

Knowledge required by the OHS professional in the US include:

- Constitutional and case law controlling safety, health, and the environment.
- Operational procedures to plan/develop safe work practices.
- Safety, health and environmental sciences.
- Design of hazard control systems (i.e. fall protection, scaffoldings).
- Design of recordkeeping systems that take collection into account, as well as storage, interpretation, and dissemination.
- Mathematics and statistics.
- Processes and systems for attaining safety through design.

Some skills required by the OHS professional in the US include (but are not limited to):

- Understanding and relating to systems, policies and rules.
- Holding checks and having control methods for possible hazardous exposures.
- Mathematical and statistical analysis.
- Examining manufacturing hazards.
- Planning safe work practices for systems, facilities, and equipment.
- Understanding and using safety, health, and environmental science information for the improvement of procedures.
- Interpersonal communication skills.

Differences between Countries and Regions

Because different countries take different approaches to ensuring occupational safety and health, areas of OSH need and focus also vary between countries and regions. Similar to the findings of the ENHSPO survey conducted in Australia, the Institute of Occupational Medicine in the UK found that there is a need to put greater emphasis on work-related illness in the UK. In contrast, in Australia and the US, a major responsibility of the OHS professional is to keep company directors and managers aware of the issues that they face in regards to occupational health and safety principles and legislation. However, in some other areas of Europe, it is precisely this which has been lacking: "Nearly half of senior managers and company directors do not have an up-to-date understanding of their health and safety-related duties and responsibilities."

Identifying Safety and Health Hazards

Hazards, Risks and Outcomes

The terminology used in OSH varies between countries, but generally speaking:

- A hazard is something that can cause harm if not controlled.
- The outcome is the harm that results from an uncontrolled hazard.
- A risk is a combination of the probability that a particular outcome will occur and the severity of the harm involved.

"Hazard", "risk", and "outcome" are used in other fields to describe e.g. environmental damage, or damage to equipment. However, in the context of OSH, "harm" generally describes the direct or indirect degradation, temporary or permanent, of the physical, mental, or social well-being of workers. For example, repetitively carrying out manual handling of heavy objects is a hazard. The outcome could be a musculoskeletal disorder (MSD) or an acute back or joint injury. The risk can be expressed numerically (e.g. a 0.5 or 50/50 chance of the outcome occurring during a year), in relative terms (e.g. "high/medium/low"), or with a multi-dimensional classification scheme (e.g. situation-specific risks).

Hazard Identification

Hazard identification or assessment is an important step in the overall risk assessment and risk management process. It is where individual work hazards are identified, assessed and controlled/eliminated as close to source (location of the hazard) as reasonably as possible. As technology, resources, social expectation or regulatory requirements change, hazard analysis focuses controls more closely toward the source of the hazard. Thus hazard control is a dynamic program of prevention. Hazard-based programs also have the advantage of not assigning or implying there are "acceptable risks" in the workplace. A hazard-based program may not be able to eliminate all risks, but neither does it accept "satisfactory" – but still risky – outcomes. And as those who calculate and manage the risk are usually managers while those exposed to the risks are a different group, workers, a hazard-based approach can by-pass conflict inherent in a risk-based approach.

The information that needs to be gathered from sources should apply to the specific type of work from which the hazards can come from. As mentioned previously, examples of these sources include interviews with people who have worked in the field of the hazard, history and analysis of past incidents, and official reports of work and the hazards encountered. Of these, the personnel interviews may be the most critical in identifying undocumented practices, events, releases, hazards and other relevant information. Once the information is gathered from a collection of sources, it is recommended for these to be digitally archived (to allow for quick searching) and to have a physical set of the same information in order for it to be more accessible. One innovative way to display the complex historical hazard information is with a historical hazards identification map, which distills the hazard information into an easy to use graphical format.

Risk Assessment

Modern occupational safety and health legislation usually demands that a risk assessment be carried out prior to making an intervention. It should be kept in mind that risk management requires risk to be managed to a level which is as low as is reasonably practical.

This assessment should:

- Identify the hazards
- Identify all affected by the hazard and how
- Evaluate the risk
- Identify and prioritize appropriate control measures

The calculation of risk is based on the likelihood or probability of the harm being realized and the severity of the consequences. This can be expressed mathematically as a quantitative assessment (by assigning low, medium and high likelihood and severity with integers and multiplying them to obtain a risk factor), or qualitatively as a description of the circumstances by which the harm could arise.

The assessment should be recorded and reviewed periodically and whenever there is a significant change to work practices. The assessment should include practical recommendations to control the risk. Once recommended controls are implemented, the risk should be re-calculated to determine if it has been lowered to an acceptable level. Generally speaking, newly introduced controls should lower risk by one level, i.e., from high to medium or from medium to low.

Contemporary Developments

On an international scale, the World Health Organization (WHO) and the International Labour Organization (ILO) have begun focusing on labour environments in developing nations with projects such as Healthy Cities. Many of these developing countries are stuck in a situation in which their relative lack of resources to invest in OSH leads to increased costs due to work-related illnesses and accidents. As a 2007 Factsheet from the European Agency for Safety and Health at Work states: "Countries with less developed OSH systems spend a far higher percentage of GDP on work-related injury and illness – taking resources away from more productive activities. The ILO estimates that work-related illness and accidents cost up to 10% of GDP in Latin America, compared with just 2.6% to 3.8% in the EU." There is continued use of asbestos, a notorious hazard, in some developing countries. So asbestos-related disease is, sadly, expected to continue to be a significant problem well into the future.

Nanotechnology

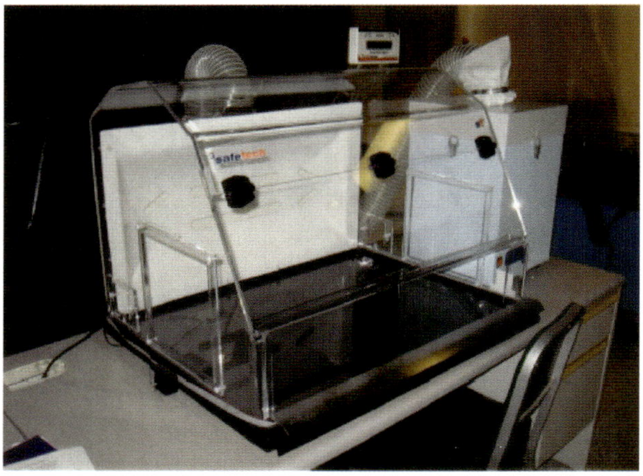

A nanomaterial containment hood, an example of an engineering control used to protect workers handling them on a regular basis.

Nanotechnology is an example of a new, relatively unstudied technology. A Swiss survey of one hundred thirty eight companies using or producing nanoparticulate matter in 2006 resulted in forty completed questionnaires. Sixty five per cent of respondent companies stated they did not have a formal risk assessment process for dealing with nanoparticulate matter. Nanotechnology already presents new issues for OSH professionals that will only become more difficult as nanostructures become more complex. The size of the particles renders most containment and personal protective equipment ineffective. The toxicology values for macro sized industrial substances are rendered inaccurate due to the unique nature of nanoparticulate matter. As nanoparticulate matter decreases in size its relative surface area increases dramatically, increasing any catalytic effect or chemical reactivity substantially versus the known value for the macro substance. This presents a new set of challenges in the near future to rethink contemporary measures to safeguard the health and welfare of employees against a nanoparticulate substance that most conventional controls have not been designed to manage.

Occupational Health Disparities

Occupational health disparities refer to differences in occupational injuries and illnesses that are closely linked with demographic, social, cultural, economic, and/or political factors.

Occupational Stress

Occupational stress consists of the harmful physical and psychological consequences to individuals that result when an imbalance exists between demands of the work environment and individual needs, abilities, and resources. Most people experience some level of occupational stress on occasion, and it is generally believed that such short-lived, episodic experiences do not pose serious or lasting harm to the individual. However, when a significant level of occupational stress persists for an extended period of time, potentially serious physical and psychological harm may occur. Although occupational stress is most often considered to be undesirable, the notion of good stress, also referred to as eustress, has been used to describe stress that motivates and energizes the worker to learn new skills and perform more effectively, without the debilitating impact typically associated with occupational stress. In the case of good stress, the result of having successfully mastered the challenge posed by the stressful condition is a sense of personal satisfaction and achievement.

Occupational stress in the American workplace is widespread and appears to be growing. When asked about primary sources of stress in their lives, a substantial number of people point to conditions in their work environments such as long hours, workload, poor communication, management problems, and lack of support, to name several examples. Recent surveys of workers suggest that as many as 80% of workers experience stress in their work, and 40% of respondents find their work to be very or extremely stressful. Occupational stress poses a very real threat to the quality of life for employees within an organization as well as a serious threat to the productivity and profitability of the organization itself.

Gender

Frustrated Man at A Desk

Men and women are exposed to many of the same stressors. Although men and women might not differ in overall strains, women are more likely to experience psychological distress, whereas men experience more physical strain. Desmarais and Alksnis suggest two explanations for the greater psychological distress of women. First, the genders may differ in their awareness of negative feelings, leading women to express and report strains, whereas men deny and inhibit such feelings. Second, the demands to balance work and family result in more overall stressors for a woman that leads to increased strain.

The Kenexa Research Institute released a global survey of almost 30,000 workers which showed that females suffered more workplace distress than their male counterparts. According to the survey, women's stress level were 10% higher for those in supervisory positions, 8% higher stress in service and production jobs than men, and 6% higher in middle and upper management than men in the same position.

Causes

Job stress results from various interactions of the worker and the environment of the work they perform their duties. Location, gender, environment, and many other factors contribute to the buildup of stress. Job stress results from the interaction of the worker and the conditions of work. Views differ on the importance of worker characteristics versus working conditions as the primary cause of job stress. The differing viewpoints suggest different ways to prevent stress at work. Differences in individual characteristics such as personality and coping skills can be very important in predicting whether certain job conditions will result in stress. In other words, what is stressful for one person may not be a problem for someone else. This viewpoint underlies prevention strategies that focus on workers and ways to help them cope with demanding job conditions. In general, occupational stress is caused by a mismatch between perceived effort and perceived reward, and/or a sense of low control in a job with high demands. Low social support at work and job insecurity can also increase occupational stress. Psychosocial stressors are a major cause of occupational stress.

Working Conditions

Although the importance of individual differences cannot be ignored, scientific evidence suggests that certain working conditions are stressful to most people. Such evidence argues for a greater emphasis on working conditions as the key source of job stress, and for job redesign as a primary prevention strategy. Large surveys of working conditions, including conditions recognized as risk factors for job stress, were conducted in member states of the European Union in 1990, 1995, and 2000. Results showed a time trend suggesting an increase in work intensity. In 1990, the percentage of workers reporting that they worked at high speeds at least one-quarter of their working time was 48%, increasing to 54% in 1995 and to 56% in 2000. Similarly, 50% of workers reported they work against tight deadlines at least one-fourth of their working time in 1990, increasing to 56% in 1995 and 60% in 2000. However, no change was noted in the period 1995–2000 (data not collected in 1990) in the percentage of workers reporting sufficient time to complete tasks.

Workload

In an occupational setting, dealing with workload can be stressful and serve as a stressor for employees. There are three aspects of workload that can be stressful.

- Quantitative workload or overload: Having more work to do than can be accomplished comfortably.
- Qualitative workload: Having work that is too difficult.
- Underload: Having work that fails to use a worker›s skills and abilities.

Workload as a work demand is a major component of the demand-control model of stress. This model suggests that jobs with high demands can be stressful, especially when the individual has low control over the job. In other words, control serves as a buffer or protective factor when demands or workload is high. This model was expanded into the demand-control-support model that suggests that the combination of high control and high social support at work buffers the effects of high demands.

As a work demand, workload is also relevant to the job demands-resources model of stress that suggests that jobs are stressful when demands (e.g., workload) exceed the individual›s resources to deal with them.

Long Hours

A substantial percentage of Americans work very long hours. By one estimate, more than 26% of men and more than 11% of women worked 50 hours per week or more in

2000. These figures represent a considerable increase over the previous three decades, especially for women. According to the Department of Labor, there has been a rise in the number of hours in the work place by employed women, an increase in extended work weeks (>40 hours) by men, and a considerable increase in combined working hours among working couples, particularly couples with young children.

Evidence of occupational stress due to an individual's status in the workplace

Status

A person's status in the workplace can also affect levels of stress. While workplace stress has the potential to affect employees of all categories; those who have very little influence to those who make major decisions for the company. However, less powerful employees (that is, those who have less control over their jobs) are more likely to suffer stress than powerful workers. Managers as well as other kinds of workers are vulnerable to work overload.

Economic Factors

Economic factors that employees are facing in the 21st century have been linked to increased stress levels. Researchers and social commentators have pointed out that the computer and communications revolutions have made companies more efficient and productive than ever before. This boon in productivity however, has caused higher expectations and greater competition, putting more stress on the employee.

The following economic factors may lead to workplace stress:

- Pressure from investors, who can quickly withdraw their money from company stocks.

- The lack of trade and professional unions in the workplace.
- Inter-company rivalries caused by the efforts of companies to compete globally.
- The willingness of companies to swiftly lay off workers to cope with changing business environments.

Bullying

Bullying in the workplace can also contribute to stress. This can be broken down into five different categories:

- Threat to profession status
- Threat to personal status
- Isolation
- Excess work
- Destabilization i.e. lack of credit for work, meaningless tasks etc.

This in effect can create a hostile work environment for the employees that, which in turn, can affect their work ethic and contribution to the organization.

Narcissism and Psychopathy

Thomas suggests that there tends to be a higher level of stress with people who work or interact with a narcissist, which in turn increases absenteeism and staff turnover. Body finds the same dynamic where there is corporate psychopath in the organization.

Workplace Conflict

Interpersonal conflict among people at work has been shown to be one of the most frequently noted stressors for employees. Conflict has been noted to be an indicator of the broader concept of workplace harassment. It relates to other stressors that might co-occur, such as role conflict, role ambiguity, and workload. It also relates to strains such as anxiety, depression, physical symptoms, and low levels of job satisfaction.

Sexual Harassment

Women are more likely than men to experience sexual harassment, especially for those working in traditionally masculine occupations. In addition, a study indicated that sexual harassment negatively affects workers' psychological well-being. Another study found that level of harassment at workplaces lead to differences in performance of work related tasks. High levels of harassment were related to the worst outcomes, and no harassment was related to least negative outcomes. In other words, women who had experienced a higher level of harassment were more likely to perform poorly at workplaces.

Occupational Group

Lower occupational groups are at higher risk of work-related ill health than higher occupational groups. This is in part due to adverse work and employment conditions. Furthermore, such conditions have greater effects on ill-health to those in lower socio-economic positions.

Individual and Organizational Perspectives on Occupational Stress

Efforts to explore the origin of occupational stress have approached the question from the perspectives of both the individual and the workplace environment. Obviously, characteristics of the individual and the work environment interact to produce the various individual and organizational outcomes observed in studies of occupational stress.

Individual Perspective

When individuals experience stress, the brain mobilizes the body's systems for defensive action in what is commonly known as the fight-or-flight response. This response is characterized by an increase in various bodily hormones that stimulate an increase in several vital functions such as heart rate, respiration, blood pressure, and muscle tension. This fight-or-flight response plays a key role in humans' ability to protect themselves when in potentially dangerous situations, either by fleeing or fighting. However, if the stressful situation persists unresolved, the prolonged arousal of vital functions may eventually deplete the body's resources, contributing to injury or disease.

Symptoms

According to the National Institute for Occupational Safety and Health (NIOSH), a federal agency responsible for research and policy on matters of work-related illness and injury, several symptoms have been identified as early warning signs of occupational stress. These symptoms may include headaches, difficulty concentrating, sleep disturbances, upset stomach, mood disturbances, job dissatisfaction, and disturbed relationships. These early warning signs are fairly easy to recognize. Of greater concern, however, is the fact that prolonged occupational stress can contribute, in concert with other factors, to more serious physical and mental health problems that are chronic in nature and potentially life threatening. NIOSH has identified three broad categories of health problems believed to be associated with prolonged stress that include cardiovascular disease (e.g., heart disease, high blood pressure), musculoskeletal disorders (e.g., tendonitis, lower back pain), and psychological disorders (e.g., anxiety, depression).

In addition to physical and psychological difficulties, it has also been shown that individuals experience stress as a result of competing work and family roles, that is, work-family conflict. With the increase in the number of women in the workforce, the growing number of dual-worker families, and the changing nature of men's and women's roles in American society, it is not surprising that a significant number of people report difficulties managing the often competing demands associated with their work and family roles. Many employees with families report feeling torn between expectations of them in the workplace and their responsibilities as spouses and parents. This conflict can further contribute to the level of stress experienced.

Influencing Factors

It has been well established that individual characteristics have a significant impact on an individual's perception, experience, and management of stress. Two individuals employed in the same work setting may provide very different reports of their own occupational stress as a result of their unique perspectives, preferences, needs, values, abilities, and experience. Perspectives may be so divergent that one person experiences stress as a positive phenomenon, serving to arouse and motivate strong performance or signify the need for a change in work or lifestyle. Another person may experience stress as negative and disruptive of performance. Often, both perspectives prove true; that is, stress may be viewed as positive and functional in the short term, but, over time, prolonged stress becomes deleterious to the individual as it weakens, erodes, and undermines psychological and physical health.

The nature and extent of occupational stress experienced, as well as one's response to it, appears to be influenced by several different personality characteristics. Not surprisingly, individuals who are highly idealistic, set unrealistic goals, overidentify with others, and have a high need for self-affirmation tend to demonstrate greater vulnerability to occupational stress. Individuals who have an internal locus of control (tend to attribute outcomes to their own efforts rather than to factors outside of their control such as luck or chance) tend to be somewhat less susceptible to occupational stress than those with an external locus of control. Self-efficacy, the belief that one has the ability to respond effectively to the demands of the work situation, appears to buffer against the negative impact of stress. Finally, an active coping style, or one in which the individual faces challenges directly and proactively, has been associated with lower levels of occupational stress as well as with more effective response strategies for managing existing stress.

Intervention

When the focus of occupational stress is the individual, the emphasis of intervention is on helping the individual to manage the stress being experienced in a way that minimizes or eliminates the negative impact of stress on work performance and the individual's health. Many companies provide stress management assistance to their employees as a benefit of employment. This assistance might include psychoeducational programs designed to teach workers about occupational stress and its potential effects on life and health as well as strategies for reducing work-related stress. Such strategies might also include training in specific behavioral management techniques such as time management and relaxation, for example. Employee assistance programs (EAPs) are another form of help that is frequently provided on-site by employing organizations. EAPs are typically staffed by professionals with training in mental health service delivery and offer individual counseling for both work-related and personal problems. According to NIOSH, although interventions focused on stress management are often relatively inexpensive and effective in reducing symptoms of occupational stress, the benefits are often short-lived. Furthermore, because the problem of occupational stress is so widespread, interventions that target employees more broadly have the potential for greater impact. Finally, interventions focusing on the individual often ignore or overlook the real sources of the problem that exist in the work environment. For these reasons, interventions that are targeted at the work environment are often preferable.

Organizational Perspective

Organizations are complex systems structured to accomplish specific tasks (e.g., manufacture a product, provide a service) and to provide control and accountability. Although modern theories regarding organizations tend to discourage rigid, hierarchical, bureaucratic designs, these characteristics are true of most current organizations, at least to some degree. Many of these features of organizations, as well as the environments within which they operate, are sources of stress for employees who work in them. The symptoms of occupational stress may be found in the behavior of employees or in patterns within the organization's structure.

Symptoms

Many different symptoms point to stress in the work environment. High levels of job dissatisfaction and related concerns are almost certain to arise eventually in stressful environments. Chronic absenteeism and tardiness are often common in highly stressful work environments, too. An unusually high turnover rate among employees is another common symptom of occupational stress that can be quite costly to the organization in terms of the time and resources required to train new people. Negative social behaviors, such as aggression, may occur more frequently in highly stressful environments. The term going postal, coined after several incidents in which disgruntled postal employees shot and killed fellow workers, has come into popular use to refer to workplace anger and violence. Increasing frequencies of stress-related physical and psychological problems among employees contribute to substantial increases in healthcare costs. Declines in productivity and the quality of workmanship are also common manifestations of a highly stressful work environment. Left unresolved, these symptoms can have devastating effects on the organization and the people employed there.

Influencing Factors

It is generally agreed that work environments have become increasingly more stressful over time. Downsizing (cutting positions or services to save costs), outsourcing (delegating jobs or services to external specialists to cut costs), and the growth of non-permanent employment contracts are examples of changes that have contributed substantially to making employees' work lives more stressful. Many workers believe their jobs are less secure than before and feel as though they have less control over their work lives. One of the most prominent theories of job stress hypothesizes that job demands interact with control to influence the amount of stress likely to be experienced. For example, high job demands are likely to be experienced as more stressful when employees have less control over their work than when they have more control.

Organizations are often described as complex human systems, composed of several important subsystems that are highly interdependent. These subsystems, although referred to by various names, exist within all organizations and compose the organization's structure. Three commonly identified subsystems include the operational (includes communication, roles, authority, power, norms, reward systems), purposive (includes goals, objectives, mission, core values), and methodological (includes technology, procedures, materials used) subsystems. Not surprisingly, different types and sizes of organizations operate most effectively with different types of structures. When given elements of an organization's structure are poorly designed or managed, the result can be considerable workplace stress. For example, expectations for employees with specialized skills to serve in multiple roles may result in excessively heavy workloads and long hours for those individuals. Likewise, excessive noise or overcrowding in the workplace may occur as a result of specific technology or procedures being used. Both of these examples illustrate how structural elements of the organization can produce stressful conditions for workers. If such conditions are permitted to persist over time, it is likely that symptoms of stress will eventually emerge both in the work environment (e.g., through reduced productivity) and in the physical and psychological health of employees.

Leadership style is a very important element of organizational structure that can greatly impact workplace stress. Leaders can influence the nature and quality of the work environment in several important ways. For example, leaders can either encourage or discourage open communication across individuals and groups within the organization. Rigid, bureaucratic leaders typically tend to discourage employees from communicating effectively with one another (horizontally) and with the leaders themselves (vertically). Poor communication may inhibit the quantity and quality of information that is available and considered in decision making within the organization. Potentially negative outcomes from poorly informed decisions may further contribute to workplace stress.

Intervention

Given the magnitude and breadth of the problem of occupational stress, it is generally believed that interventions that focus on changing the structure of the organization (work environment) will produce more effective and more enduring outcomes than interventions focusing on individual employees. In light of the fact that so much of work-related stress can be attributed to problems in organizational structure, it makes good sense to design organizations in ways that facilitate strong work performance while minimizing the stress experienced by employees of the organization. In the past, it was largely assumed that stress was an inevitable companion to organizational productivity. However, considerable evidence has emerged to suggest that the opposite is true; that is, productivity and worker satisfaction will be higher in organizational environments that optimize the health and welfare of employees.

Remedial approaches that focus on treating individual casualties of the stress produced in work environments will always be necessary and appropriate. Thus, there will be a continuing place for procedures designed to foster more effective management of stress at the individual level. However, the most desirable approach is one that is preventive in nature. Such an approach will require creating and maintaining work environments that are highly sensitive and responsive to the needs of workers. In reality, the most successful efforts to addressing workplace stress will likely incorporate both individual and organizational interventions.

There are many different models for facilitating organizational change, but they tend to share several common elements. For example, those to be affected by future change(s) ought to be included in all aspects of the change process. This kind of inclusion fosters ownership of the change, increases the likelihood that the change will meet the needs of those affected by it, and reduces the likelihood of resistance due to perceptions that change is being imposed. Collaboration among affected parties in identifying problems and designing and implementing changes will increase the likelihood that plans are based upon sound information and diverse perspectives. The feelings and preferences of individuals must be considered along with factual data as the worth and dignity of the individual is of paramount importance.

Measurement of Occupational Stress

Stress is measured through ASSET, a new Organizational Screening Tool (OST) which is the advanced form of the well-established and extensively used Occupational Stress Indicator (OSI). ASSET is a sorter and applicable to all occupations has been developed and has been used successfully in health care organizations with adequate evidence of construct and discriminated validity, which is a very effective tool in diagnosing occupational stress, combining both the sources and the effects of stress.

ASSET conceptualizes occupational stress as influenced by a variety of sources, such as work relationships, work-life balance, overload, job security, control, resources and communication, pay and benefits, as well as an evaluation of the employee's perception of the potential sources of stress that relate to the fundamental nature of the job itself named Aspects of the Job. ASSET divides organizational commitment in two subscales as follows:

- Commitment of the Organization to the Employee (COE), and
- Commitment of the Employee to the Organization (CEO).

High score in both scales indicates increased commitment. The former measures the extent to which individuals feel that their organization is committed to them, on the other hand the latter measures the degree that employees feel loyal and committed to the organization.

Theories of Occupational or Work-Related Stress

Contemporary theories of stress have moved away from understanding the construct as either a response or as an external event, but rather view it as a dynamic interaction between the individual and their environment. Contemporary theories of stress, either explicitly or implicitly, recognize the central role of psychological processes (such as perception, cognition and emotion) in understanding:

- How the individual recognizes, experiences, and responds to stressful situations.
- How they attempt to cope with that experience.
- How it might affect their physical, psychological and social health).

There are several key contemporary theories in the scientific literature that have helped to clarify the causes and mechanisms that underpin work-related stress. Many of these theories have been extensively researched and have been used to guide approaches to intervention. The first three are structural model and they describe the key variables and interactions among those variables in relation to outcomes of interest. The fourth is a process model which describes the mechanisms that underpin the relationship between the antecedent and outcomes. Four prominent theories are presented here.

Person-Environment Fit theory (P-E Fit theory)

Much of contemporary stress theory finds its origins in the early work of the social science research group at the University of Michigan and in particular the work of Kahn, French, Caplan and van Harrison. Together they developed the Person-Environment (P-E) Fit theory. P-E Fit theory argues that stress can arises due to a lack of fit between the individual's skills, resources and abilities, on the one hand, and the demands of the work environment, on the other hand.

The P-E Fit theory makes explicit the interaction between the individual and the environment in shaping their response to work situations and events, but also highlights the importance of the individual's perception of the environment; and the interaction between them. Logically, this lack of fit can take three forms:

1. The demands of the work environment exceed the employee's ability.
2. The employee's needs consistently fail to be met by the work environment.
3. A combination of the two situations exists (i.E., Where an employee's needs are not being met while at the same time their abilities are over-stretched).

Job Demand-Control (Support) Theory

The Job Demand-Control (JCD) model and its expanded version the Job Demand-Control-Support model have dominated the field of occupational stress research for more than two decades. The JCD model postulates that job strain results from the interaction between two dimensions of the work environment: psychological job demands and job control.

Psychological demands traditionally referred to workload, operated mainly in terms of time pressure and role conflict. However, more recently, cognitive and emotional demands and interpersonal conflict dimensions define the contemporary construct of psychological demand. Job control (also often also referred to decision latitude in the literature) refers to the person's ability to control their work activities, and is defined by two key components: (a) decision authority (worker's ability to make decisions about their job); and (b) skill discretion (the breadth of skills used by the worker on the job). The JCD theory suggests that individuals experiencing high demands paired with low control are more likely to experience psychological strain, work-related stress, and, in the long term, poor physical and mental health.

The model was later extended to include a social dimension: social support. The JCDS model postulates that social support can moderate the negative impact of job strain on worker's physical and mental health. This model suggests that the most at-risk group of poor physical and mental health are those workers who are exposed to job strain (high demands and low control) paired with low workplace support (a phenomenon referred to as iso-strain.

Effort-Reward Imbalance Model (ERI model)

The ERI model was developed by Johannes in the early 1990's. This theory assumes that effort at work is spent as part of a psychological contract, based on the norm of social reciprocity, where effort spent at work is paired with rewards provided in terms of money, esteem, career opportunities. An imbalance (non-reciprocal) relationship between the effort spent and rewards received can result in the emotional distress associated with a stress response, and an increased risk of ill-health. Siegrist suggests that stress related to the imbalance between effort and rewards can arise under three conditions: namely,

- Has a poorly defined work contract or where the employee has little choice concerning alternative employment opportunities.
- Accepts the imbalance for reasons such as the prospect of improved working conditions.
- Copes with the demands at work through over commitment.

Transactional Model

Transaction models build upon the interaction between the individual and their environment, but provide an additional focus on the underlying psychological and physiological mechanisms which underpin the overall process. Cox and MacKay suggested that stress is the result of a dynamic interaction between the individual and the environment. However, unlike previous models of stress, central to this model is the individual's cognitive assessment of the perceived demands made on the worker, and their perceived capability, skills and resources to deal with those demands. That is, stress results when the perceived demands outweigh the perceived capability of the workers. What an individual finds or perceives to be stressful can vary both between and within individuals, and can differ over occasions and time. In this way, any aspect of the work environment can be perceived as a stressor, and therefore unlike previous models transactional models are not limited by the types and number of psychosocial hazards they can account for. The cognitive assessment by the worker of the perceived demands and capabilities can be influenced by a number of factors: personality, situational demands, coping skills, pervious experiences, and any current stress state already experienced. In addition, this model acknowledges that stress can manifest physiologically, psychologically, behaviorally and socially with detrimental consequences to both the individual and the organization. Research indicates that the relationship between psychosocial hazards and health outcomes is mediated by a variety of factors; the transactional model accounts for the complex relationship by acknowledging individual variation and differences in the stress process.

Occupational Stress Consequences

The following health disorders in humans are associated with occupational stress:

- Cardiovascular diseases.
- Anxiety disorders, depression, post-traumatic stress syndrome.
- Gastric ulcers.
- Metabolism disorders.
- Skeletal and muscular complaints.

Several Risk Behavior Is Often Characteristic of A Person In Stress:

- Smoking.
- Excess use of alcohol.
- Using drugs.
- Little movement and sedentary lifestyle.
- Unhealthy diet.

A clear determination of the occupational stress is complicated because stress and related damage depends on several circumstances. A situation that creates an uncomfortable tension on one person may be stimulating for another; a task which is manageable on one day could seem overwhelming due to some personal problems, or a conflict between employees which influences the feelings and work performance. Therefore, frank communication between the employer and employee is crucial. If the employee senses a strong pressure that triggers fatigue and other negative phenomena, the employer must be informed so that the employer could adapt work accordingly. If the employee does not wish to talk to the employer directly, they can turn to the working environment representative.

The following problems related to the high stress level occur in the organization:

- Low creativity, work ability and commitment of employees.
- High level of incapability to work.
- Great staff turnover.
- Increased risk of occupational accidents.
- Quality problems that generate costs emanating from the dissatisfaction of customers.

Occupational stress more broadly also influences the society:

- An employee in stress could take it out on their family and co-citizens, increasing the aggressiveness in society.
- The society's productiveness suffers.
- The national health insurance system must bear the majority of costs related to the occupational stress.
- If the employee is not capable to return to work due to stress, the pension system bears costs.

Prevention

A combination of organizational change and stress management is often the most useful approach for preventing stress at work. Both organizations and employees can employ strategies at organizational and individual levels. Generally, organizational level strategies include job procedure modification and employee assistance programs (EAP). Individual level strategies include taking vacation. Getting a realistic job preview to understand the normal workload and schedules of the job will also help people to identify whether or not the job fit them.

How an organization can prevent job stress:

- Ensure that the workload is in line with workers' capabilities and resources.
- Design jobs to provide meaning, stimulation, and opportunities for workers to use their skills.
- Clearly define workers' roles and responsibilities.
- To reduce workplace stress, managers may monitor the workload given out to the employees. Also while they are being trained they should let employees understand and be notified of stress awareness.
- Give workers opportunities to participate in decisions and actions affecting their jobs.
- Improve communications-reduce uncertainty about career development and future employment prospects.
- Provide opportunities for social interaction among workers.
- Establish work schedules that are compatible with demands and responsibilities outside the job.
- Combat workplace discrimination (based on race, gender, national origin, religion or language).
- Bringing in an objective outsider such as a consultant to suggest a fresh approach to persistent problems.
- Introducing a participative leadership style to involve as many subordinates as possible to resolve stress-producing problems.
- Encourage work-life balance through family-friendly benefits and policies.

An insurance company conducted several studies on the effects of stress prevention programs in hospital settings. Program activities included (1) employee and management education on job stress, (2) changes in hospital policies and procedures to reduce organizational sources of stress, and (3) the establishment of employee assistance programs. In one study, the frequency of medication errors declined by 50% after prevention activities were implemented in a 700-bed hospital. In a second study, there was a 70% reduction in malpractice claims in 22 hospitals that implemented stress prevention activities. In contrast, there was no reduction in claims in a matched group of 22 hospitals that did not implement stress prevention activities.

Telecommuting is another way organizations can help reduce stress for their workers. Employees defined telecommuting as "an alternative work arrangement in which employees perform tasks elsewhere that are normally done in a primary or central workplace, for at least some portion of their work schedule, using electronic media to interact with others inside and outside the organization." One reason that telecommuting gets such high marks is that it allows employees more control over how they do their work. Telecommuters reported more job satisfaction and less desire to find a new job. Employees that worked from home also had less stress, improved work/life balance and higher performance rating by their managers.

A systematic review of stress-reduction techniques among healthcare workers found that cognitive behavioral training lowered emotional exhaustion and feelings of lack of personal accomplishment.

An Example to Understand it Better

As like other professions teachers have been leaving their profession at the start of the 21st century, as the most cause problems with rising stress. Most alarming is that 30% of all beginner teachers are leaving their profession and annual teacher turnover is higher when compared with turnover of all other occupations. A report in June 2002 issue of the National Association of Secondary School Principles of the USA showed that annual turnover of professions other than teaching averaged 11% while annual turnover for the teaching profession was 15.7%. If stress is very low then the performance of work is also very low, for the medium stress performance of work is very high and for the high stress performance of work is also very low expressed that the executive as well as the teachers experienced a moderate level of stress, the executives experienced more stress than the teachers realized.

Teachers' occupational stressor means their stress source which changes stress intensity related to teaching or teaching context of the individual. There are three aspects of it as follows:

1. The first aspect deals with school organizational factor, such as mission requirements, role demands, role ambiguity, role confliction, work overload and organization change.
2. The second involves individual factor, for example, individual ability, individual expectation, character feature and relationship among the groups.
3. The third relates to environmental factor, i.e., uncertainty of economics, politics and technology.

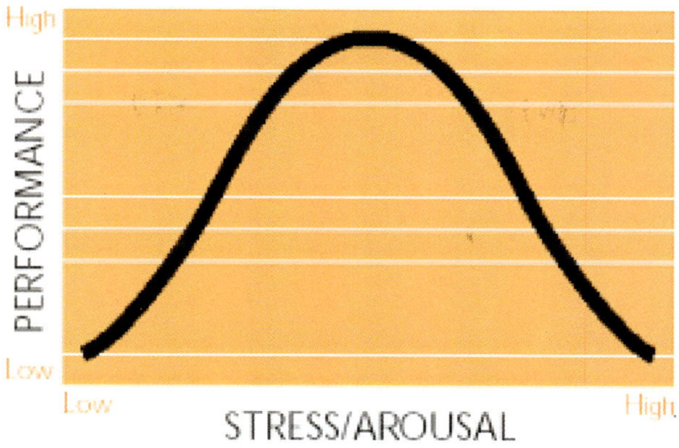

Figure: Relation between stress and performance of work

The common stressors teachers experience frequently worldwide are as follows:

- Poor classroom working conditions i.E., Small room but many students.
- Job insecurity.
- Teachers bring work at home daily and no time to finish at work.
- Lack of adequate instructional materials and teaching resources.
- Low salaries.
- Little opportunity for promotions.
- Annual incremental raises not in pace with cost of living.
- Heavy workload gives teachers no time to relax within a day.
- Physical and mental abuse by students.
- Conflict between amount of time to teach and curriculum.
- Responsible to administration or board objectives and policies.
- Lack of proper facilities within a school such as heating or cooling, lighting, media facilities, noise, room structure etc.

Teacher Burnout

Burnout, a concept coined during the 1970's, is a concern across many disciplines and of international interest. Maslach, a noted researcher on burnout and the author of the Maslach Burnout Inventory, defines burnout as "a syndrome of emotional exhaustion and cynicism that occurs frequently among individuals who do 'people-work' of some kind. A key aspect of the burnout syndrome is increased feelings of emotional exhaustion." Burnout is depression, feelings of alienation, helplessness and hopelessness, loss of idealism and spirit, and physical and emotional drain. Burnout is a prolonged response to chronic emotional and interpersonal stressors on the occupation, and is defined by the three dimensions of exhaustion, cynicism, and inefficacy. Teachers are leaving their profession and are joining in government jobs and private firms, some are seeking early retirement, and others are simply dropping out. A great number of teachers have laid down their pointers and chalk largely globally because of decreased funding, limited personal control over their teaching, and lack of societal commitment. Burnout is a more serious problem to the profession than job change or early retirement because it renders a teacher unable to cope, although he or she remains in the classroom. This is due to limited resources, long hours, marginal working conditions, and often unreasonable demands from those receiving services, lead to chronic stress, and ultimately, burnout. Those who experience burnout typically find that they are tired, drained, and without enthusiasm.

A psychological model of how stress leads to burnout describes it as a syndrome resulting from teachers' inability to protect themselves against threats to their self esteem and well being. When the coping mechanisms fail to stem the demands then stress increases and threatens the teachers' mental and physical well-being ultimately leading to teachers quitting or burning out express that "Work appears as a major source of stress for working people, teachers appear to experience more stress through work than non-teachers". The average length of a teaching career in the USA is now down to eleven years and one quarter of all beginning teachers leave teaching within four years.

Coping Strategies of Stress

If any teacher feels stress seriously he must take steps to cope them. There are some ways of coping of stress as follows:

- Identify co-workers, friends, and family members which are helpful and can talk to them and need to make social plans with them. Sometimes campus-wide social activities can reduce the stress.
- Practice deep breathing, visualization, meditation, yoga, stretching or some other techniques which helps to relax.
- Take a break from whenever is causing the stress and take parts in laugh, play, sing, fun and praying.
- Take a healthy balance diet and enough sleep. Exercise moderately at least two to three times a week. Remember that caffeine, alcohol, cigarettes and drugs do not help solve your problems and actually contribute to stress.
- Arrange a picnic for mental refreshment which reduces stress and creates good teaching environment in the campus.
- Teachers' working conditions should be improved by providing some social facilities such as satellite TV, computers with Internet connections.

- The school curriculum should be restructured to include both teachers-centered and students-centered co-curricular activities as regards to inter-house sporting activities in the schools.

Risk Management at Workplace

Many definitions of risk and alternative approaches to risk management exist but there is no universally accepted definition of risk. A risk can be conceptualized as having two basic elements, one relating to the probability that an event, decision or activity will have undesirable negative outcomes, and the other relating to the severity of those outcomes as follows:

Risk = Probability of occurrence of an event × Consequences for a given time period. The level of risk or risk factor associated with the likely negative effects of a given stressor can be calculated as follows:

$$\text{Risk factor, } R_F = E_s \times C_s,$$

where, E_s = the perceived level of the exposure (stressor) and C_s = consequences (the correlation between stressor and stress outcome). Values of C_s are obtained by calculating the coefficient of correlation between the stressors and stress outcomes, r and then converting r into r^2, and then converting the value of r^2 into percent i.e., $C_s = r^2 \times 100$. McFarlane conducted a stress audit within a leading United Kingdom (UK) retail organization. He calculated the perceived level of the exposure, E_s = 9.85 and the correlation between workload and mental health, $r = 0.09$ and $r^2 = 0.0081$ so that $C_s = r^2 \times 100 = 0.81$, and hence by ($R_F = E_s \times C_s$) the risk factor can be calculated as follows:

$$RF = E_s \times C_s = 9.85 \times 0.81 \approx 8.$$

The decision as to the acceptability of a risk is determined by considering the cost of lowering the risk level, against the benefits from a reduction in risk.

Normative risk factors for an industry can be calculated, where the norm for exposure, E_{norm} and for consequences, C_{norm} is known as:

$$RF_{norm} = E_{norm} \times C_{norm}.$$

Calculated RF_{norm} from an illustrative data as follows: The norm Pressure Management Indicator score = 12.46, so that, E_{norm} =12.46 the correlation between workload and the General Health Questionnaire (GHQ) is r = 0.05, r^2 = 0.0025, $C_{norm} = r^2 \times 100 = 0.25$ and hence by ($RF_{norm} = E_{norm} \times C_{norm}$) the risk norm factor can be calculated as follows:

$$RF_{norm} = 12.46 \times 0.25 \approx 3,$$

where the GHQ is a well-validated instrument that measures symptoms of mental ill-health and PMI measures the level of perceived stress, can be used to obtain E_{norm} for a particular sample.

Work Accident

An occupational accident is a sudden and unexpected accident caused by external factors due to which the employee suffers injuries. As an occupational accident is considered an accident occurring in the work place premises or, on the way home from work or vice versa or while the employee is away on a business trip or running an errand ordered by the employer.

The occupational accident is severe if the employee dies or if he or she suffers a permanent or severe injury. Severe injuries are for example:

- Long bone fractures or fractures which may need surgery, difficult spinal fractures, viscerocranial fractures, multiple rib fractures and hemothorax, severe skull compound fractures, severe fractures to neck, chest or lumbar spine.
- Severe dislocation causing permanent injury.
- Loss of limb for example a finger or part of it, shortening of limb.
- Injury to the abdominal cavity organs requiring surgery.
- Brain injury leaving also light injuries.
- Loss or permanent weakening of speech, sight or hearing, loss of eyes.
- Burn or similar injuries to skin requiring skin transplantation, extensive burns or frostbites.
- Severely disfiguring deformities, for example loss of ear lobe or nose or part of it or other severe deformity.
- Permanent severe health harm or a life-threatening disease or lesion.

When the final severeness of the injury is not clear, the accident must be reported, as the nature of the injuries may become clear only through later examinations.

Serious injuries must also be reported to the police, who must conduct a police investigation at the scene of the accident.

Neglecting to report an accident to the Regional State Administrative Agency and the police is a penal offence as well as not reporting all accident to the insurance company.

Common Causes of Workplace Accidents

Shortcuts

Humans are notoriously lazy, so taking shortcuts is a rather common practice in all walks of life, not necessarily work alone. However, when workers take shortcuts at work, especially when they are working around dangerous machinery or lethal chemicals, they are only exposing themselves to a potential catastrophe. Simply put, shortcuts that are taken on the job are not actually shortcuts. They are simply increasing your risk of injury, or worse, death.

Overconfidence

Confidence is always a great thing to have, but there is also such a thing as too much confidence. When workers walk into work everyday with the attitude that, "It will never happen to me", they are setting an attitude that leads to incorrect procedures, methods, and tools while working. Be confident, but remember that you are not invincible.

Poor, or Lack of Housekeeping

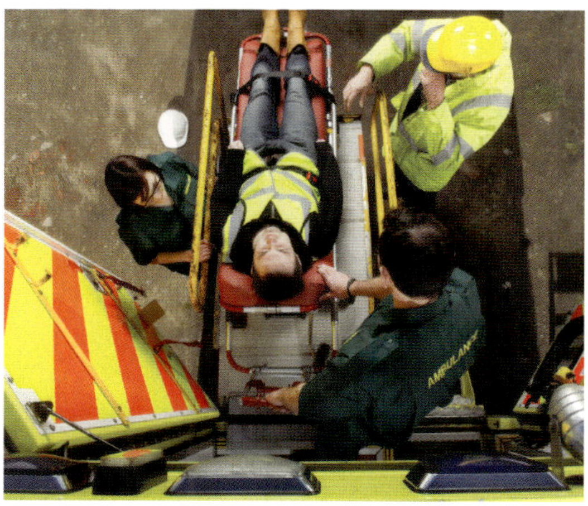

Starting a Task Before Getting All Necessary Information

Whenever someone walks through your workplace, they can get a pretty good idea of your attitude towards workplace safety by just looking at how well you've kept up your area. Housekeeping is one of the most accurate indicators of the company's attitude towards production, quality, and worker safety. A poorly kept up area leads to hazards and threats everywhere. Not only does good housekeeping lead to heightened safety, but it also sets a good standard for everyone else in the workplace to follow.

The quickest way to get a job done is to do it right the first time. To do it right the first time, you need to make sure that you have any and all pertinent information relating to the task you will be performing. Workers, who begin a job with just half the information, or half the instructions, are essentially doing the job while blind. Remember this; it's not stupid to ask questions, it is stupid not to.

Neglecting Safety Procedures

This is probably the worst thing that any employee at any level in the organization can do. Deliberately neglecting set safety procedures in the workplace doesn't just endanger yourself, but it endangers the workers around you as well as the company as a whole. Casually following safety procedures doesn't work either. You are paid to follow workplace safety procedures, not your own.

Mental Distractions

Everyone has a life outside of the workplace, and sometimes life can take dips and turns that affect your emotions and your mood negatively. However, as harsh as it sounds, workers cannot let mental distractions from their personal lives affect their performance at work. Not only will they become less aware of their surroundings and less safe, but they will also become less productive, costing the company time and money.

Lack of Preparation

You may have heard of something called Job Hazard Analysis (JHA). JHA's are an effective method of figuring out the best way to work safely and efficiently. When workers begin a task without thinking through the process beforehand, or hastily start without any type of planning, they are setting themselves up for failure. The employee should make sure to plan his work, then to work his plan.

Types of Workplace Injuries

1. Overexertion Injuries: This includes injuries related to pulling, lifting, pushing, holding, carrying, and throwing activities at work. Overexertion not only consistently been the number one workplace injury according but is also the most expensive.

2. Slipping/Tripping: The number 2 cause of workplace injuries, this pertains to falls on wet and slippery floors or trips over something lying on the floor. Attention to what

is going on around them is required of employees and employers will have safety guidelines to ensure spills are promptly cleaned and no debris is present which can be dangerous.

3. Falling from Heights: This type of fall happens from an elevated area such as roofs, ladders, and stairways. They can be caused by slip and fall accidents or due to faulty equipment. These types of accidents can be reduced by the use of proper personal protection gear, training and employee diligence.

4. Reaction Injuries: These are injuries caused by slipping and tripping without falling. These incidents can cause muscle injuries, body trauma, and a variety of other medical issues. It can be hard to prevent such incidents but it is important for employees to pay attention to what is going on in the environment around them.

5. Falling Object Injuries: Objects that fall from shelves or dropped by another person can cause very serious injuries. Head injuries are a common result of this type of accident. Employee›s diligence and employer focus on keeping the work environment free from hazards are key to preventing these types of injuries. Of course, proper personal protection gear usage, such as a hard hat, can be instrumental in keeping the employee safe.

6. Walking Into Injuries: This happens when a person accidentally runs into concrete objects such as walls, doors, cabinets, glass windows, table, chairs etc. Head, knee, neck, and foot injuries are common results. Employee›s diligence and employer focus on keeping the work environment free from hazards are key to preventing these types of injuries.

7. Vehicle Accidents: Employees who drive for business purposes are often injured in auto accidents, some of which can be fatal. Employee Safe-Driver training and employer safe driving policies are likely to reduce accidents.

8. Machine Entanglement: This type of injury usually occurs in a factory where heavy equipment and machinery are used. Clothing, shoes, fingers and hair are by everyday equipment when no precaution is taken. Protective equipment and attention to personal details are necessary to avoid these incidents.

9. Repetitive Motion Injuries: This type of workplace injury is one of those less obvious but definitely harmful ones in the long run. Repetitive motions such as typing and using the computer 24/7 can strain muscles and tendons causing back pain, vision problems, and carpal tunnel syndrome. Employee training and the use of proper ergonomic equipment can help keep these incidents low.

10. On the Job Violent Acts: Attacks caused by office politics and other arguments have led to serious physical injuries. Workplace violence employee training and employee diligence in watching out for suspicious activities can help keep these incidents at bay.

Occupational Injury

An occupational injury is defined as any personal injury, disease or death resulting from an occupational accident; an occupational injury is therefore distinct from an occupational disease, which is a disease contracted as a result of an exposure over a period of time to risk factors arising from work activity.

An occupational accident is an unexpected and unplanned occurrence, including acts of violence, arising out of or in connection with work which results in one or more workers incurring a personal injury, disease or death.

A case of occupational injury is the case of one worker incurring an occupational injury as a result of one occupational accident. An occupational injury could be fatal (as a result of occupational accidents and where death occurred within one year of the day of the accident) or non-fatal with lost work time.

Incapacity for work is the inability of the victim, due to an occupational injury, to perform the normal duties of work in the job or post occupied at the time of the occupational accident. Incapacity can be permanent or temporary. Cases of permanent incapacity for work are cases of occupational injury where the persons injured were never able to perform again the normal duties of work in the job or post occupied at the time of the occupational accident causing the injury. Cases of temporary incapacity are cases of occupational injury where the workers injured were unable to work from the day after the day of the accident, but were later able to perform again the normal duties of work in the job or post occupied at the time of the occupational accident causing the injury within a period of one year from the day of the accident.

It has been estimated that worldwide there are more than 350,000 workplace fatalities and more than 270 million workplace injuries annually. In 2000 there were approximately 2.9 billion workers worldwide. Occupational injuries resulted in the loss of 3.5 years of healthy life for every 1,000 workers. 300,000 of the occupational injuries resulted in a fatality.

The most common occupations associated with these hazards vary throughout the world depending on the major industries in a particular country. Overall, the most hazardous occupations are in farming, fishing, and forestry. In more developed countries, construction and manufacturing occupations are associated with high rates of spine, hand, and wrist injuries.

Occupation Injuries Statistics for Different Countries

In the United States in 2012, 4,383 workers died from job injuries, 92% of which were men, and nearly 3 million nonfatal workplace injuries & illness were reported which cost businesses a collective loss of $198.2 billion and 60 million workdays. In 2007, 5,488 workers died from job injuries, 92% of which were men, and 49,000 died from work-related injuries. NIOSH estimates that 4 million workers in the U.S. in 2007 suffered from non-fatal work related injuries or illnesses.

According to data from the National Institute for Occupational Safety and Health (NIOSH) and the Bureau of Labor Statistics, an average of 15 workers die from traumatic injuries each day in the United States, and an additional 200 workers are hospitalized.

In a study in the state of Washington, injured workers were followed for 14 years to determine the long term effects of work injury on employment. The work injuries resulted in an average of 1.06 years of lost productivity for each of the 31,588 allowed claims.

Dangerous Sectors

In the U.S. the Bureau of Labor Statistics makes available extensive statistics on workplace accidents and injuries. For example:

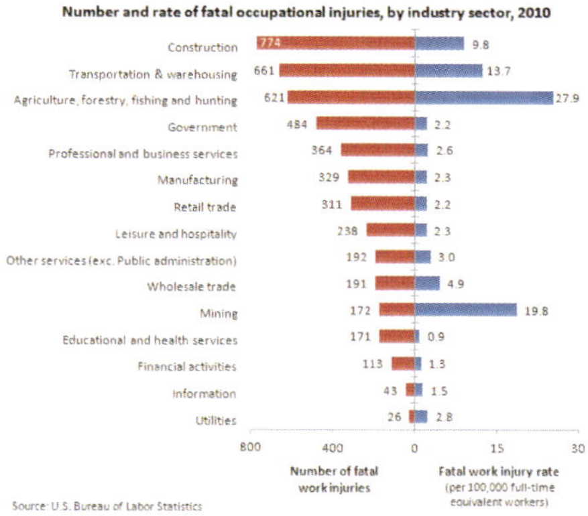

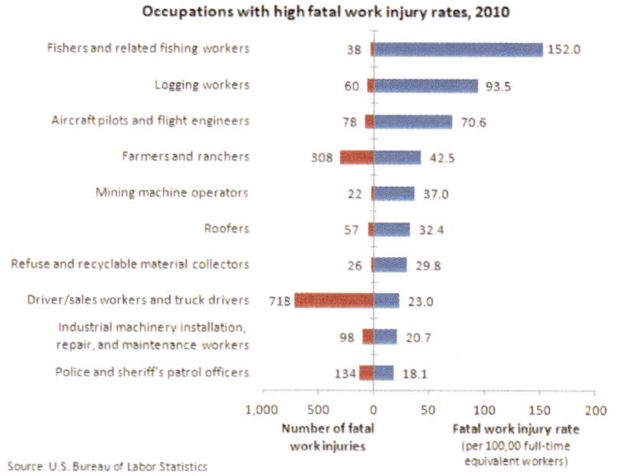

Common Injuries

As in the United Kingdom, slips, trips and falls are common and account for 20-40% of disabling occupational injuries. Often these accidents result in a back injury that can persist to a permanent disability. In the United States, a high risk of back injuries occurs in the health care industry. 25% of reported injuries in health care workers in the state of Pennsylvania are for back pain. Among nurses, the prevalence of lower back pain may be as high as 72% mostly as a result of transferring patients. Fortunately, some of these injuries can be prevented with the availability of patient lifts, improved worker training, and allocation of more time to perform work procedures. Another common type of injury is carpal tunnel syndrome associated with overuse of the hands and wrists. Studies on a cohort of newly hired workers have thus far identified forceful gripping, repetitive lifting of > 1 kg, and using vibrating power tools as high risk work activities.

Additionally, noise exposure in the workplace can cause hearing loss, which accounted for 14% of reported occupational illnesses in 2007. Many initiatives have been created to prevent this common workplace injury. For example, the Buy Quiet program encourages employers to purchase tools and machines that produce less noise and the Safe-In-Sound Award was created to recognize companies and program that excel in the area of hearing loss prevention.

Accidental injection or needle stick injuries are a common injury that plagues agriculture workers and veterinarians. The majority of these injuries is located to the hands or legs, and can result in mild to severe reactions, including possible hospitalization. Due to the wide variety of biologics used in animal agriculture, needle stick injuries can result in bacterial or fungal infections, lacerations, local inflammation, vaccine/antibiotic reactions, amputations, miscarriage, and death. Due to daily human-animal interactions, livestock related injuries are also a prevalent injury of agriculture workers, and are responsible for the majoriy of nonfatal worker injuries on dairy farms. Additionally, approximately 30 people die of cattle and horse-related deaths in the United States annually.

Employees At Risk

Perhaps the most important personal factor that predisposes to an increased risk is age. In the United States in 1998 17 million workers were over age 55 and by 2018 this population is expected to more than double. Workers in this age group are more likely to suffer from lower back pain that may be worsened by work conditions that normally do not affect a younger worker. Older workers are also more likely by be killed in a construction related fall. They are also at higher risk for injury due to age-related hearing loss, visual impairment, and use of multiple prescription medications that has been linked to higher rates of work injuries. In addition to age, other personal risk factors for injury include obesity particularly its associated risk with back injury, and depression.

Lack of proper education or training can also predispose an individual to an occupational injury. For example, there is limited needle stick injury awareness among agriculture workers, and there is a need for comprehensive programs to prevent needle stick injuries on livestock operations. Proper animal handling techniques and training, or stockmanship, can also decrease the risk of livestock injury. A handler's timing, positioning, speed, direction of movement, and sounds made will affect the behavior of an animal and consequently the safety of the handler. The agriculture industry has begun to focus more on proper education and training, and has made a variety of resources available to producers. For example, organizations like the Upper Midwest Agriculture Safety and Health Center (UMASH) have a variety of informational fact sheets and training videos easily accessible online. Additionally, organizations like Beef Quality Assurance offer stockmanship training seminars and demonstrations.

Regulation

In the United States, the Occupational Safety and Health Administration (OSHA) sets and enforces national standards for occupational safety across all sectors.

In the United Kingdom in 2013/2014, 133 people were killed at work. Of those 133 people, 89 were employed, while 44 were self-employed. In 2013/2014, an estimated 629,000 injuries occurred at work. Of these injuries 629,000 injuries, 203,000 led to more than 3 days absence from work. Of these, over 148,000 resulted in the victim being absent from work for more than 7 days.

Dangerous Sectors

- Construction: 42 deaths
- Agriculture: 27 deaths
- Waste and Recycling: four deaths
- Other: 60 deaths

Of all the workplace accidents that resulted in death, the most common were falls from height, contact with moving machinery and being struck by a vehicle. These types of accidents resulted in over half of all recorded deaths.

Common Injuries

Slips, trips and falls account for over a third of all injuries that happen at work. Incorrect handling of items was the most common cause of injuries that led to absences from work of more than 7 days. Upper limb injuries represented 47.3% of workplace injuries in 2010-2011, the most common area injured.

In all, over 1,900,000 working days were lost in 2013/2014 due to slips, trips and falls.

Employees At Risk

Unsurprisingly, occupation is the biggest influence on the risk of workplace injuries. Workers new to the job are at a much higher risk of injury than more experienced staff, while shift workers and part-time staff also have a greater risk of being injured at work.

Employer Prosecutions

The Health & Safety Executive (HSE) prosecuted 582 cases in 2013/2014, with at least one conviction secured in 547 cases (94%).

Local authorities prosecuted a total of 92 cases during the same period, with at least one conviction achieved in 89 cases (97%).

A total of 13,790 notices were issued by the HSE and local authorities, with over £16,700,000 issued in fines.

In Taiwan, there were 14,261 occupational injuries recorded in 2010. 45% of these involved trauma to the upper limbs.

Prevention

There are many methods of preventing or reducing industrial injuries, including anticipation of problems by risk assessment, safety training, control banding, personal protective equipment safety guards, mechanisms on machinery, and safety barriers. In addition, past problems can be analyzed to find their root causes by using a technique called root cause analysis. A 2013 Cochrane review found low-quality evidence showing that inspections, especially focused inspections, can reduce work-related injuries in the long term.

Occupational Noise

Occupational noise is the amount of acoustic energy received by an employee's auditory system when they are working in the industry. Occupational noise, or industrial noise, is often a term used in occupational safety and health, as sustained exposure can cause permanent hearing damage.

"Twenty-two million workers are exposed to potentially damaging noise at work each year. Last year, U.S. business paid more than $1.5 million in penalties for not protecting workers from noise." - OSHA

Occupational noise is considered an occupational hazard traditionally linked to loud industries such as ship-building, mining, railroad work, welding, and construction, but can be present in any workplace where hazardous noise is present.

Regulation

In the United States, the National Institute for Occupational Safety and Health (NIOSH) and the Occupational Safety and Health Administration (OSHA) work together to provide standards and regulations for noise in the workplace.

National Institute for Occupational Safety and Health (NIOSH), Occupational Safety and Health Administration (OSHA), Mine Safety and Health Administration (MSHA), Federal Railroad Administration (FRA) have all set standards on hazardous occupational noise in their respective industries. Each industry is different, as workers' tasks and equipment differ, but most regulations agree that noise becomes hazardous when it exceeds 85 decibels, for an 8-hour time exposure (typical work shift). This relationship between allotted noise level and exposure time is known as an Exposure action value (EAV) or Permissible exposure limit (PEL). The EAV or PEL can be seen as equations which manipulate the allotted exposure time according to the intensity of the industrial noise. This equation works as an inverse, exponential, relationship. As the industrial noise intensity increases, the allotted exposure tie, to still remain safe, decreases.

The above calculations of PEL and EAV are based on measurements taken to determine the intensity of that particular industrial noise. A-weighted measurements are commonly used to determine noise levels that can cause harm to the human ear. There are also special exposure meters available that integrate noise over a period of time to give an Leq value (equivalent sound pressure level), defined by standards.

Risks of Occupational Hearing Loss

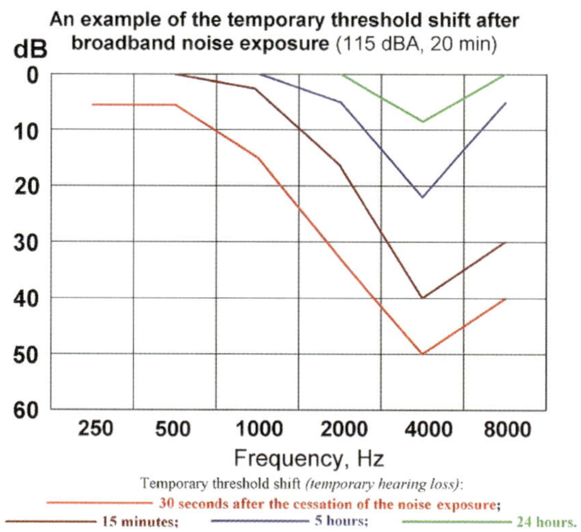

Temporary hearing loss after loud noise exposure. If such exposure is long enough, this temporary threshold shift may become permanent.

Occupational noise, if experienced repeatedly, at high intensity, for an extended period of time, can cause noise-induced hearing loss (NIHL) which is then classified as occupational hearing loss.

Noise, in the context of industrial noise, is hazardous to a person's hearing because of its loud intensity through repeated long-term exposure. In order for noise to cause hearing impairment for the worker, the noise has to be close enough, loud enough, and sustained long enough to damage the hair cells in the auditory system. Occupational hearing loss or Noise-induced hearing loss for more information regarding the physiology of hearing loss. These factors have been taken into account by the governing occupational health and safety organization to determine the unsafe noise exposure levels and durations for their respective industries.

Noise can also affect the safety of the employee and others. Noise can be a causal factor in work accidents as it may mask hazards and warning signals and impede concentration. High intensity noise interferes with vital workplace communication which increases the chance of accidents and decreases productivity.

Noise may also act synergistically with other hazards to increase the risk of harm to workers. In particular, toxic materials (e.g. some solvents, metals, asphyxiants and pesticides) have some ototoxic properties that may affect hearing function.

Modern thinking in occupational safety and health further identifies noise as hazardous to workers' safety and health. This hazard is experienced in various places of employment and through a variety of sources.

Reduction

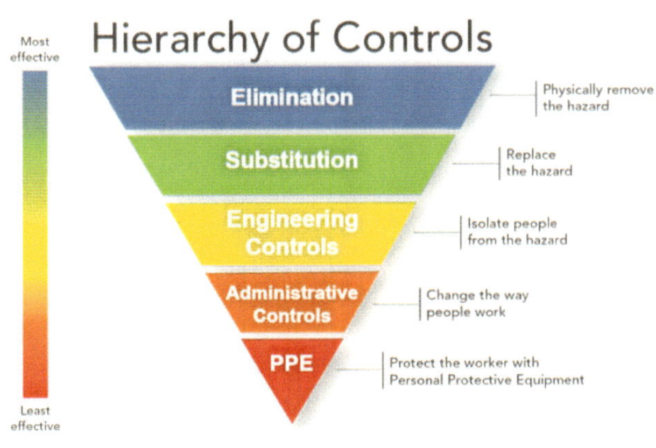

Hierarchy of Controls Guideline

There are several ways to limit your exposure to hazardous occupational noise. The hierarchy of controls is a guideline for reducing hazardous noise. First, the company can eliminate the noise source. If the noise source cannot be eliminated, the company must try to reduce the noise with alternative methods. This process is called acoustic quieting.

Acoustic quieting is the process of making machinery quieter by damping vibrations to prevent them from reaching the observer. The company can isolate the certain piece of machinery by placing materials on the machine or in between the machine and the worker to decreases the signal intensity that reaches the worker's ear.

To decrease an employee's exposure to hazardous noise, the company can also take administrative control by limiting the employee's exposure time. This can be done by changing work shifts and switching employees out from the noise exposure area. Lastly, to decrease occupational noise exposure, hearing protection should be used. There are several types of earplugs and earmuffs that can be used to attenuate the noise to a safe level.

Initiatives

Since the hazards of occupational noise exposure were realized, programs and initiatives such as the US Buy Quiet program have been set up to regulate or discourage noise exposure. The Buy Quiet initiative promotes the purchase of quieter tools and equipment and encourages manufacturers to design quieter machines. Additionally, the Safe-In-Sound Award was created to recognize successes in hearing loss prevention programs or initiatives.

Occupational Heat Stress

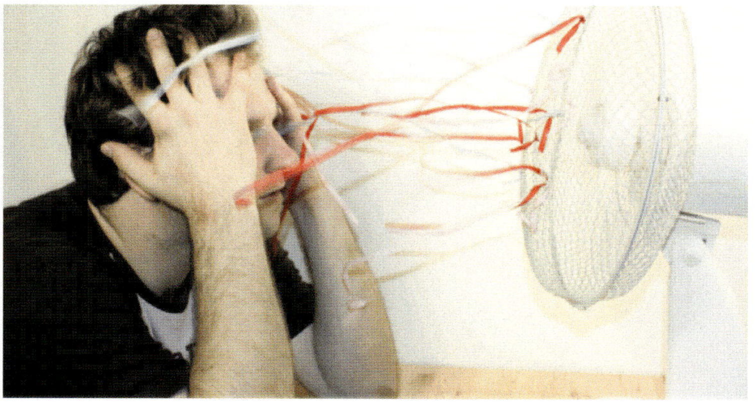

Occupational heat exposure threatens the health of a worker not only when heat illness occurs but also when a worker's performance and work capacity is impaired. Occupational contexts that involve hot and humid climatic conditions, heavy physical workloads and/or protective clothing create a strenuous and potentially dangerous thermal load for a worker.

Workers who are exposed to extreme heat or work in hot environments may be at risk of heat stress. Exposure to extreme heat can result in occupational illnesses and injuries. Heat stress can result in heat stroke, heat exhaustion, heat cramps, or heat rashes. Heat can also increase the risk of injuries in workers as it may result in sweaty palms, fogged-up safety glasses, and dizziness. Burns may also occur as a result of accidental contact with hot surfaces or steam.

Workers at risk of heat stress include outdoor workers and workers in hot environments such as firefighters, bakery workers, farmers, construction workers, miners, boiler room workers, factory workers, and others. Workers at greater risk of heat stress include those who are 65 years of age or older, are overweight, have heart disease or high blood pressure, or take medications that may be affected by extreme heat.

Prevention of heat stress in workers is important. Employers should provide training to workers so they understand what heat stress is, how it affects their health and safety, and how it can be prevented.

Types of Heat-related Illnesses

Heat Stroke

Heat stroke is the most serious heat-related illness. It occurs when the body becomes unable to control its temperature: the body's temperature rises rapidly, the sweating mechanism fails, and the body is unable to cool down. When heat stroke occurs, the body temperature can rise to 106°F or higher within 10 to 15 minutes. Heat stroke can cause death or permanent disability if emergency treatment is not given.

Symptoms

Symptoms of heat stroke include:

- Confusion, altered mental status, slurred speech.
- Loss of consciousness (coma).
- Hot, dry skin or profuse sweating.
- Seizures.
- Very high body temperature.
- Fatal if treatment delayed.

First Aid

Take the following steps to treat a worker with heat stroke:

- Call 911 for emergency medical care.
- Stay with worker until emergency medical services arrive.
- Move the worker to a shaded, cool area and remove outer clothing.
- Cool the worker quickly with a cold water or ice bath if possible; wet the skin, place cold wet clothes on skin, or soak clothing with cool water.
- Circulate the air around the worker to speed cooling.

Place cold wet clothes or ice on head, neck, armpits, and groin; or soak the clothing with cool water.

Heat Exhaustion

Heat exhaustion is the body's response to an excessive loss of the water and salt, usually through excessive sweating. Workers most prone to heat exhaustion are those that are elderly, have high blood pressure, and those working in a hot environment.

Symptoms

Symptoms of heat exhaustion include:

- Headache
- Nausea
- Dizziness
- Weakness
- Irritability
- Thirst
- Heavy sweating
- Elevated body temperature
- Decreased urine output.

First Aid

Treat a worker suffering from heat exhaustion with the following:

- Take worker to a clinic or emergency room for medical evaluation and treatment.
- If medical care is unavailable, call 911.
- Someone should stay with worker until help arrives.

- Remove worker from hot area and give liquids to drink.
- Remove unnecessary clothing, including shoes and socks.
- Cool the worker with cold compresses or have the worker wash head, face, and neck with cold water.
- Encourage frequent sips of cool water.

Rhabdomyolysis

Rhabdomyolysis is a medical condition associated with heat stress and prolonged physical exertion, resulting in the rapid breakdown, rupture, and death of muscle. When muscle tissue dies, electrolytes and large proteins are released into the bloodstream that can cause irregular heart rhythms and seizures, and damage the kidneys.

Symptoms

Symptoms of rhabdomyolysis include:

- Muscle cramps/pain.
- Abnormally dark (tea or cola colored) urine.
- Weakness.
- Exercise intolerance.
- Asymptomatic.

First Aid

Workers with symptoms of rhabdomyolysis should:

- Stop activity.
- Increase oral hydration (water preferred).
- Seek immediate care at the nearest medical facility.
- Ask to be checked for rhabdomyolysis (i.e., blood sample analyzed for creatine kinase).

Heat Syncope

Heat syncope is a fainting (syncope) episode or dizziness that usually occurs with prolonged standing or sudden rising from a sitting or lying position. Factors that may contribute to heat syncope include dehydration and lack of acclimatization.

Symptoms

Symptoms of heat syncope include:

- Fainting (short duration).
- Dizziness.
- Light-headedness during prolonged standing or suddenly rising from a sitting or lying position.

First Aid

Workers with heat syncope should:

- Sit or lie down in a cool place.
- Slowly drink water, clear juice, or a sports drink.

Heat Cramps

Heat cramps usually affect workers who sweat a lot during strenuous activity. This sweating depletes the body's salt and moisture levels. Low salt levels in muscles causes painful cramps. Heat cramps may also be a symptom of heat exhaustion.

Symptoms

Muscle cramps, pain, or spasms in the abdomen, arms, or legs.

First Aid

Workers with heat cramps should:

- Drink water and have a snack and/or carbohydrate-electrolyte replacement liquid (e.g., sports drinks) every 15 to 20 minutes.
- Avoid salt tablets.
- Get medical help if the worker has heart problems, is on a low sodium diet, or if cramps do not subside within 1 hour.

Heat Rash

Heat rash is a skin irritation caused by excessive sweating during hot, humid weather.

Symptoms

Symptoms of heat rash include:

- Looks like red cluster of pimples or small blisters.
- Usually appears on the neck, upper chest, groin, under the breasts, and in elbow creases.

First Aid

Workers experiencing heat rash should:

- When possible, a cooler, less humid work environment is best treatment.
- Keep rash area dry.
- Powder may be applied to increase comfort.
- Ointments and creams should not be used.

Examples of High Risk Occupations

Workers in many occupations are at high risk for exposure to heat stress. Some of the higher risk occupations include firefighter, bakery worker, miner, military personnel, construction worker, factory worker, boiler room worker, landscaper, some athletes and agricultural worker.

Symptoms of Heat Stress

The main symptoms of heat stress are perspiration, increased heart rate, and dehydration. Other general symptoms include painful muscle cramps, extreme weakness, nausea, dizziness, headache, breathing fast and clammy, pale, cool, and/or moist skin or red, dry skin.

Prevention

Employers can establish prevention programs, which focus on having protocols to gradually increases workloads and concede on allowing on more breaks for new hired workers. Employers can control heat stress through engineering controls, work practices, providing training, implementing an acclimatization schedule, providing water and encouraging workers to drink often, and ensuring workers take appropriate rest breaks to cool down.

Occupational Standards

International Organization for Standardization helps set standards for monitoring environments, analyzing data, and interpreting results.

Section 5(a)(1) of the Occupational Safety and Health Act of 1970 declares that “Each employer shall furnish to each of his employees employment and a place of employment which are free from recognized hazards that are causing or are likely to cause death or serious physical harm to his employees.”

The Mine Safety and Health Administration provides guidelines and recommendations to employers for preventing heat stress among workers. There guidelines and recommendations are not enforced regulations, but instead completely voluntary.

National Institute for Occupational Safety and Health (NIOSH) conducts research on occupational hazards such as heat stress in order to provide better intervention methods and protect workers.

Only three of the fifty states have created worker regulations regarding heat: California, Washington, and Minnesota. California Code of Regulations states that employers of high risk outdoor workers are entitled to protection against heat. The employer must provide access to water and shade, practice high heat procedures, practice emergency response procedures, and practice acclimatization methods. Washington State Legislature states that employers of high risk outdoor workers follow regulations to prevent heat stress. Minnesota Administrative Rules state that indoor ventilation and temperature are regulated to prevent heat stress.

Recommended Standards

Beginning in 1972, NIOSH published a recommended standard for hot work environments, and has periodically revised to take new scientific findings into account. The intent of the NIOSH Recommended Standard for Occupational Exposure to Heat and Hot Environments is to prevent injury, disease, death, and reduced productivity. The recommendations include workplace limits and surveillance, medical monitoring, surveillance of heat-related sentinel health events, posting of hazardous areas, protective clothing and equipment, worker information and training, control of heat stress, and record keeping.

"Protect Your Workers from Heat Stress", CDC

Control of Heat Stress

Control of heat stress has recommended general requirements, engineering controls, work and hygienic practices, and a heat alert program.

General Requirements

NIOSH recommends that every employer should create and implement a written program aimed at reducing heat exposures. Engineering and work practice controls should be used to reduce exposures, and a heat alert program should be implemented.

Engineering Controls

Air temperatures should be reduced so it does not exceed skin temperatures. Radiant heat should be reduced by creating barriers around the source. Evaporative heat loss can be increased by increasing air movement around the worker.

Work and Hygienic Practices

The time workers spend in hot environments should be limited, with an increase of recovery time spent in cool environments. Use of more efficient procedures and tools is beneficial to reducing metabolic demands of the job. Heat tolerance may be increased by implementing a heat tolerance plan and increasing physical fitness. Employees should be trained to recognize and treat the early signs and symptoms of heat illnesses, and employers should provide cool water for employees.

Heat Alert Program

Heat alert programs should be developed for implementation when hotter than normal temperatures, or a heat wave occurs.

Occupational Dust Exposure

Occupational dust exposure can occur in various settings, including agriculture, forestry, and mining. Dust hazards include those that arise from handling grain and cotton, as well as from mining coal. Wood dust, commonly referred to as "sawdust", is another occupational dust hazard that can pose a risk to workers' health.

Without proper safety precautions, dust exposure can lead to occupational lung diseases.

Occupational Dust

Dust particles are generated by the disturbance/agitation of rock/mineral, dry grain, timber, or fiber material. The particles generated can range in size from 1μm to 100μm, and can "become airborne depending on their origin, physical characteristics and ambient conditions."

Occupational Dust Examples

Types of dust present in the occupational setting include:

- Rock/mineral dusts
- Metallic dusts
- Chemical dusts
- Grain and produce dusts
- Molds and spores

Occupations and Dust Hazards

Mining

During various mining procedures in which rock/minerals are broken up and collected for processing, mineral dusts are created and become airborne. Inhalation of these dusts can lead to various respiratory illnesses, depending on the dust type (i.e. coal, silica, etc.), size of the dust particulates, and exposure duration.

Forestry

During the stages of wood processing, wood dust is generated. 'Wood dust' is "any wood particles arising from the processing or handling of woods." Sawing, routing, sanding, among other activities, form wood dust, which can then become airborne during the process of dust removal from furniture, maintenance, or equipment cleanup.

Agriculture

Dust generated from milling, handling and storage of grains can pose a threat to workers' health. During the milling process, solid agricultural grains (corn, barley, wheat, etc.) undergo crushing, grinding, or granulation. This process generates agricultural dust. Improperly handling grains can also expose workers to grain dusts. Grain storage can also present hazards to workers. Storage structures can create dangerous conditions due to gases emitted from spoiled grains and chemical fumes. "Workers may be exposed to unhealthy levels of airborne contaminants, including molds, chemical fumigants (toxic chemicals), and gases associated with decaying and fermenting silage."

Health, Safety, and Epidemiology

Exposure to occupational dusts poses many hazards to workers' health and safety. Large, airborne dust particles can obscure vision, limit mobility while on the ground, and interfere with proper machine/equipment functioning.

Characteristics of dust particles such as size and chemical qualities can determine the location and effects of the dust particles in the respiratory system.

Occupational Lung Diseases

Illnesses/Diseases that can develop due to exposure to dust in the workplace.

- Hypersensitivity Pneumonitis.
- Occupational Asthma.
- COPD (Bronchitis, Emphysema).
- Pneumoconiosis (Coal Workers' Pneumoconiosis, Asbestosis, Silicosis).
- Mycobacterial Infections.

Safety

Workplace facilities have in place safety protocol and regulations to ensure that exposure to dust/particulate matter is minimal to non-existent. Barring the elimination or substitution of the hazard, controls on the engineering and administrative levels act to protect workers from dust hazards. Personal protective equipment (PPE) is used by workers to prevent specific exposure to the hazard.

To mitigate exposure to dust in the workplace, respirators, ventilators, and eye protection are measures often employed.

Occupational Cardiovascular Disease

Occupational cardiovascular diseases are diseases of the heart or the blood vessels that are caused by occupational hazards or detrimental conditions at work. Both chemical and non-chemical risk factors can have a contribution to the incidence of occupational cardiovascular diseases. There is a strong association between cardiovascular conditions and certain toxins such as carbon monoxide, carbon disulfide and nitroglycerin, extreme temperatures, exposure to smoke, or even occupational stress and depression.

Non-Chemical Risk Factors

A 2015 SBU-report including a systematic review of non-chemical riskfactors for occupation cardiovascular disease found an association between certain occupational risk factors and developing cardiovascular disease in those:

- With mentally stressful work with a lack of control of their own working situation — with a effort-reward imbalance
- Who experience low social support at work; who experience injustice or experience insufficient opportunities for personal development; or those who experience job insecurity
- Those who work night schedules; or have long working weeks
- Those who are exposed to noise

Specifically the risk of stroke was also increased by:

- Exposure to ionizing radiation

Hypertension develops more often in those who experience job strain and who have shift-work. Differences between women and men in risk are small, however men risk suffering and dying of heart attacks or stroke twice as often as women during working life.

Chemical Risk Factors

A 2017 SBU report found evidence that workplace exposure to silica dust, engine exhaust or welding fumes is associated with heart disease. Associations also exist for exposure to arsenic, benzopyrenes, lead, dynamite, carbon disulphide, carbon monoxide, metalworking fluids and occupational exposure to tobacco smoke. Working with the electrolytic production of aluminum or the production of paper when the sulphate pulping process is used is associated with heart disease. An association was also found between heart disease and exposure to compounds which are no longer permitted in certain work environments, such as phenoxy acids containing TCDD(dioxin) or asbestos.

Workplace exposure to silica dust or asbestos is also associated with pulmonary heart disease. There is evidence that workplace exposure to lead, carbon disulphide, phenoxyacids containing TCDD, as well as working in an environment where aluminum is being electrolytically produced, is associated with stroke.

Occupational Skin Diseases

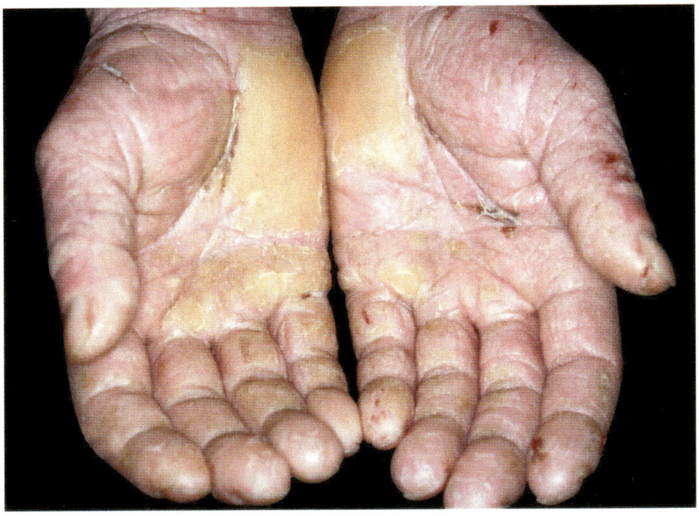

Occupational skin diseases (also known as occupational dermatoses) are skin diseases caused by exposures to hazards at work. In relation to skin the most relevant chemical hazards are irritants, sensitising, photosensitising and acnegenic agents. Biological hazards range from bacteria, fungi, viri, to skin parasites. Physical hazards include rubbing, mechanical pressure, radiations and temperature. Epidemiology data does not represent the real extent of the issue, because most of cases are not reported and the registration criteria vary from country to country too. It is estimated that 20-30% of occupational diseases are skin diseases, placing them among the three most frequent groups of occupational diseases. The most common cause of skin disease in the workplace is exposure to chemical substances. As with any occupational disease, a certain degree of causal relationship between the skin disease and the occupation must be verified during investigation. Accordingly, there are two requisites: (i) making the right and specific medical diagnosis and (ii) verifying that the skin disease is related to the occupation. For the later, it is essential to be familiar with the work activity, the hazards at work, the occupational safety and health instructions and skin protection programme at the workplace.

Skin Diseases of Chemical Origin

Exposure to chemical substances is a leading cause of occupational skin diseases. It is obvious because the worker's first line of contact with its environment, aside from the respiratory tract, is the skin. As most work involve manual activities, the skin of the hands is very likely to be involved. The gloves and clothes should be checked for their adequacy to prevent exposure, as exposure may be a source of non-dermatological occupational diseases as well. Airborne exposure may require further prevention measures, like ventilation, special protective masks and goggles.

Contact Dermatitis

The most common occupational skin disease, covering around 50-80% of cases, is irritant contact dermatitis, which is around five times more frequent than allergic contact dermatitis. It is the inflammation of the skin at the place of contact with the chemical substance. Irritant contact dermatitis and allergic contact dermatitis are undistinguishable from each other by clinical or histological examinations. Furthermore, mixed form (allergic plus irritant) contact dermatitis is also very frequent. On one hand, a substance may have both irritant and sensitising properties; in addition, the worker may be exposed to multiple substances. On the other hand, irritant exposure harms the barrier function of the skin, which promotes sensitization by enabling increased absorption of allergens. Previous or present atopic dermatitis is a strong risk factor for developing occupational contact dermatitis. Atopic dermatitis is a common skin disorder that is based on genetic disposition of decreased barrier function and impaired repair, and features itching and dryness. There is an abundant range of occupations and sectors that are at risk of occupational contact dermatitis including: hairdressers and beauticians, health care, construction, horticulture, textile industry, catering and food industry, rubber manufacturing, metalworking, electroplating and tanning.

Irritant Contact Dermatitis (ICD)

Exposure to irritant chemical substances can directly damage the skin at the place of contact and lead to ICD. Common chemical agents include acids, alkalis, organic solvents and detergents. The effect on the skin depends not only on the chemical properties and the intensity of contact, but the concentration of the chemical as well. Wet work is the leading cause of ICD. The term refers to occupational activities that require: (i) the skin of one or both hands to be in contact with water or aqueous solutions irritants; or (ii) the sustained wear of protective gloves that makes the hands moist from perspiration. Furthermore, such soakings disrupt the natural barrier in the skin and promotes the entry of sensitizing agents; thus predisposing to the development of skin allergies as well. Symptoms of ICD range from dryness, burning red rashes (erythema) and swelling (oedema) to the development of bullas (blistering), necrosis and peeling (desquamation) in more severe cases. Chronic, repeated exposures trigger the development of fully-fledged eczema: itching, cracking, crusting and flaking, oozing or even bleeding that may be present alongside the signs mentioned earlier. Patch tests are negative in ICD that has no allergic component. However, in ICD the penetration of allergens through the skin is increased, which increases the risk of developing allergy. Among atopic individuals the skin has a lower capacity to resist against irritants, thus a short exposure to low doses may trigger the symptoms of inflammation.

Skin inflammation due to the combined exposure of a chemical substance and light (phototoxic dermatitis) can lead to developing vesicles and bullas (blistering) on the inflamed skin, which is followed by a characteristic skin tan (hyperpigmentation). The most frequent chemicals involved in phototoxic dermatitis are:

- Furocoumarins (psoralens) in plants ,which may make, for example, gardeners, florist and manufacturer of cosmetic particularly vulnerable.
- Tar derivatives, which may make, for example, road construction and building insulation workers particularly vulnerable.
- Pharmaceuticals, to which exposure may be present in certain pharmaceutical plants.

Allergic Contact Dermatitis (ACD)

This form of skin inflammation may be acquired only after contact with sensitizing chemical substance. In ACD the sensitization process can take weeks, months or even years and there is no allergic reaction to be seen on the skin during that period. The hypersensitivity to the given chemical will practically last for life, although avoiding further contact with the chemical agent may lower the level of reaction. This allergy is a late onset cell-mediated immune response to the chemical substance, which explains why its symptoms have a latency period of days, which is longer than that of contact urticaria. Clinical signs of ACD are very similar to that of ICD, although itching is predominant over burning, furthermore skin lesions are not limited to the contact site but might develop at unexposed parts too. The most important occupational allergens are:

- Metals - chromium, nickel, cobalt, mercury
- p-phenylenediamine (PPD)
- Rubber additives
- Natural resins
- Artificial resins
- Biocides
- Preservatives
- Animal feeds
- Plants
- Pharmaceuticals.

A special form of ACD is photo allergic contact dermatitis. Here the chemical compound transforms into a complete sensitising agent only when exposed to ultraviolet light (mainly from the A-band spectrum). Fragrances (e.g., in cleaning, chemical industry), optical brighteners (e.g., in laundries, washing powder manufacturing), dyes (e.g., in textile industry), and pharmaceuticals are the most common chemicals involved in photo allergic contact dermatitis.

Airborne Contact Dermatitis (ABCD)

ABCD is characterized by a skin inflammation that is even and is primarily located on the face, around the eyes and neck. However, it may be present on the hands or even cloth-covered areas. This is caused by environmental dust (wood dust, textile fibers, cement, sand, glass fiber) or other chemical compound in the air (epoxy resins, acrylates, vegetable dust e.g. from the Composite/Asteraceae family, pollens, propolis). The underlying pathology of ABCD can be either ICD or ACD, where later is quiet often occupational in nature. The importance of ABCD is that it needs to be distinguished from photosensitivity, which may occur in the same parts of the body feature the same localization.

Contact Urticaria (CU)

The symptoms of CU are itching and hives (urticaria). In immunologic CU the reaction is an immunglobulin E mediated early immune response against the sensitising chemical substance. Thus in contact urticaria the chemical compound causes hives at the place of contact usually within an hour. This can be accompanied by (allergic) rhinitis, conjunctivitis, asthma and rarely anaphylaxis. There is also a non-allergic route (non-immunologic CU), where the chemical directly triggers histamine release in the skin causing urticaria. Common sources of allergic CU are proteins (latex, fish, meats, eggs), plants (crops and ornamental plants, wood), bisphenol-A, chromium and cobalt. A well-known form of allergic CU is latex-allergy among healthcare, catering and cleaning workers who are wearing latex gloves. Direct CU can be caused by stinging-nettle, preservatives, additives, dyes, fruits, vegetables, cobalt, balsam of Peru. Tackling CU is more difficult, because it requires a total elimination of the allergen due to the risk of the more serious complications; thus change of job is more frequently recommended to avoid/ prevent exposure.

Diagnosis of Contact Dermatitis

Verifying chemical allergy is usually done by (epicutaneuos) patch testing, when minute dose of diluted substances are placed onto the skin and observed periodically. If an occupational origin is suspected the general battery (baseline series), which contains everyday sensitising agents, should be supplemented by occupational series (common sensitising substances of a given sector/occupation) and/or substances used at work (based on the safety data sheets). There are further diagnostic tests available: patch test with early examination, photo patch test, use test, prick test, intracutaneous test. In vitro laboratory assays (e.g. quantitative IgE blood test) are less sensitive and specific. Tests identifying responsible immune cells are under research.

Occupational Acne

These are comedos (pinheads), papules and pustules caused mostly by industrial oils and greases. Unlike common acne, these eruptions manifest at the site of skin contact.

Tar derivatives and halogen-containing compounds (polychlorinated naphthalene, polychlorinated phenoxy phenols, 3 4-dichloroaniline and similar herbicides, iodides and bromides) may cause acne, just like certain pharmaceuticals. Car mechanics, maintenance workers are most at risk. Personal and work hygiene is of utmost importance in the prevention of oil acne. Frequent change and centralised washing of dirty work wear is necessary.

Skin Cancers Due to Chemical Exposure

Pitch, tar, soot, anthracene and compounds thereof, mineral and other oils, raw paraffin, carbazole and compounds thereof, coal tar distillation products may cause a disturbance in the epidermis. This can trigger the development of basal cell and squamous cell carcinomas, or conditions (keratotic papilloma, keratoacanthoma) that may lead to these cancers. Arsenic causes brownish pigmentation and rough, scaly patches (keratosis). Bowen's disease, squamous cell carcinoma and multiple basalioma may develop.

Prevention of Skin Diseases Due to Chemical Exposure

Proper workplace skin protection programmes adapted to the actual situation are the best way to prevent occupational skin disease. The measures follow the hierarchy of control like in other occupational safety and health programmes. The initial target is to eliminate/minimise skin contact with noxious (carcinogen, irritant and sensitising) material and wet work. Personal protective equipment, gloves should be chosen according to the (chemical) properties of the risk, taking into consideration the tasks, the work environment and the worker. In order to minimise wet work, the duration of uninterrupted wearing of gloves should be minimised as well. National initiatives and sectoral guides may provide useful information for the implementation of a skin protection plan. However, it must be always tailored to the actual situation of the given workplace. There are several examples of successful implementation of workplace skin protection plans. Information and training of workers on exposures, prevention techniques, proper use of personal protective devices and skin hygiene is of fundamental importance. Health surveillance is recommended in order to identify the shortcomings of measures and workers that need advanced protection. Support to the individual in maintaining healthy skin includes the provision of proper skin cleaning agents and skin rehydration products at the workplace.

Infectious Skin Disease

Several infectious agents may be contracted from various sources during work, including animals at work (occupational zoonoses). The diagnostic and treatment process is identical to non-occupational diseases. Diagnosis is based on the clinical signs plus history and if required on laboratory blood tests, microscopy and growing cultures. Prevention is based on the risk assessment of the given workplace that takes into account unintentional exposure to biological agents. The measures follow the hierarchy of control and include proper personal and work hygiene, the use of germicide agents and gloves. In case of zoonoses the co-operation with the veterinarian is essential.

Bacterial Infections

Occupational pyodermas (folliculitis, furuncle, carbuncle, impetigo, ecthyma, paronychia, etc.) can be caused by the Streptococcus and Staphylococcus bacteria which generate pus. These infections are common among those working in dirty environments (e.g., car mechanics, sewerage cleaners) where microtrauma (bruises, cuts) of the skin are common (e.g., butchers, slaughterhouse workers, machining of metals), or who may be in contact with infected persons (e.g., nurses, hairdressers, manicurists). Colonisation, which is the presence of microorganisms on the worker without apparent disease, and manifest infection with antibiotic resistant bacteria (e.g., MRSA) among health care workers is an emerging issue with implications for patient safety.

Erysipeloid is caused by Erysipelothrix rhusiopathiae. The infection presents in the form of marked skin inflammation that accompanies the inflammation of the underlying connective tissue (cellulitis). It is almost exclusively occupational, among butchers and workers in contact with fish and poultry.

Skin tuberculosis is a granulomatous, slowly progressing, wart-like skin lesion with regional lymph node involvement. Mycobacterium tuberculosis hominis may develop among pathologist, dissectors, and surgeons. M. tuberculosis bovis might be an occupational skin disease of vets, animal handlers, butchers, farmers. M. marinum infection is transmitted by fish and water in tanks or pools (fish tank granuloma).

Erythema chronicum migrans (ECM), which is a ring-like slowly growing red rash, is the early sign of Lyme-disease (Borrelia burgdorferi infection). It is a common occupational skin disease among forestry and horticultural workers. If untreated, later phases might feature cutaneous lymphoid hyperplasia, which is a benign skin bump, and acrodermatitis chronica atrophicans, which is characterized by initial inflammation and swelling and later atrophy of the skin at the limbs.

Fungal Infections (Mycoses)

Yeast Infections

The occupational infection, which is mainly caused by Candida albicans, occurs usually on the hands: onychomycosis (nail), paronychia (around the nail bed), interdigital mycosis (between the fingers or toes). Workers in canneries and confectioneries, and health care workers are at risk. Wearing rubber gloves and boots (wet work) and handling of sweets can be contributing factors to the development of these infections.

Dermatophyte Infections (Ringworm)

Trichophytia profunda is caused by Trichophytia verrucosum and looks like a deep bacterial skin infection. Farmers, milkers, animal handlers, and veterinarians may acquire it from infected cattle. Microsporiasis is an infection of Microsporum canis that is common among pets (dogs, cats, guinea pigs) and may cause ringworm among pet traders and breeders, vets, laboratory workers. Microsporum gypseum lives in soil and may also infect agriculture workers.

Viral Skin Diseases

Milker's nodules (Nodus mulgentium) are caused by the Paravaccinia virus. The source is the udder of the cow, although less frequently the udder of sheep or goats, where it presents as a quickly healing wound. Milkers and other animal handlers may develop the skin lesion which heals spontaneously without scarring within weeks. Orff (Ecthyma contagiosum) is caused by a Para pox virus, which is common in sheep and goats. The tender skin lesion may be accompanied by regional inflammation of the lymph vessels and nodes. The disease may be contracted by shepherds, goatherds, veterinarians and it may heal spontaneously without scarring within weeks unless the wound is superinfected.

Parasitic Skin Lesions

Although frequently unrecognized, parasitic skin diseases may have an occupational origin. Arthropod bites from animal parasites or granary mites are common in agricultural workers. Bee and wasp stings are important, because their venom may cause serious and even life-threatening allergic complications. However, most beekeepers develop a kind of resistance where even a dozen bites may not cause any skin, or other effect. Scabies is caused by Sarcoptes scabiei. This tiny mite digs itself into the skin around the wrists, elbows, hands, feet, back, buttocks, and external genitals causing itching and superficial burrows. The infection is transmitted by direct contact, e.g. during care activities. Treatment should involve contact persons without symptoms too. Control of infection is challenging for the health and social care institutions, which are most at risk.

Skin Diseases of Physical Origin

Physical exposures range from direct stimuli and thermal conditions to radiation.

Mechanical Trauma

Recurring rubbing or increased pressure can thicken the inflamed skin with "crazy-paving pattern" (lichenification) and callosity (ref). So-called occupational stigmas are frequent at body parts exposed: shoulders of sack carriers, fingertips of guitarist and violinists, knuckles of masseurs, hands of blacksmiths. The conditions are only of aesthetic nature and reversible; prevention is usually impossible.

Temperature

Besides their physiological effect on the human body, extreme temperatures can cause specific alterations in the skin. They are on the border of diseases and accidents, however, due the time characteristics, the severe forms are rather considered accidents at work. The prevention is based on insulation, appropriate clothing and work organization (rotation of workers and specified recovery areas for breaks and shelter).

Heat

Sweat stagnation can cause miliaria (~crystallina, rubra, profunda), which are small and itchy rashes (sweat rash, prickly heat). Overlapping skin surfaces can become sodden, ending up in intertrigo; which is an infectious inflammation of body folds. Heat and physical exercises may cause heat urticaria: non-allergic hives. Intense heat can cause burns. The metallurgy sector, bakeries, kitchens, summer outdoor jobs (agriculture, construction), and deep mining are typical exposures.

Cold Temperature

The effects of cold to the skin may range from progressive vasoconstriction (narrowing of blood vessels) and Raynaud-like symptoms (blanching attacks of fingers) to frostbites. Cold-storage and winter outdoor (e.g. maintenance, construction) jobs are examples for such exposure.

Ionising Radiation

Ionising radiation is used widely in medical imaging and therapy, industrial quality inspection, and the sterilization of medicines and foods. Occupational exposure to ionizing radiation, including X-ray, usually causes local skin signs (radiation dermatitis). X-ray and gamma radiation can penetrate deeper, and thus, harm inner layers than beta radiation which is absorbed superficially, or even by the clothes. As technology is improved and made safer due to consistent occupational safety and health standards, local radiation burns are very rare nowadays. The skin symptoms after a short high level exposure to ionising radiation are termed acute radio dermatitis, while the effects of long term exposure is summarized as chronic radio dermatitis. Their prevention is mainly technical (enclosure of the technology, segregation, proper maintenance) accompanied by strict protocols, shielding devices and personal protective equipment.

Acute Radio Dermatitis

After a single dose of around 10 Grays or more, local redness (erythema), swelling (oedema) and loss of skin color develops. Nausea and vomiting may be present. After days of latency the disease may progress with a painful livid inflammation with vesicles and bullas (blistering) and necrosis that heals with scar, atrophy and loss of hair.

Chronic Radio Dermatitis

Repeated exposure to smaller (3-8 Grays) doses with a total of 50-60 Grays causes atrophic indurated plaques, spider veins (telangiectasia) or even thickening (hyperkeratosis), a syndrome called termed poikiloderma. Skin cancer can arise after months or years on the grounds of poikiloderma.

Non-ionizing Radiation

Concerning occupational skin diseases, workplace exposure to ultraviolet radiation is among the most important physical risk factors. Although ultraviolet radiation is necessary for the human body to produce vitamin D, excessive exposure is harmful: it damages the skin and the eyes, causes mutation of the DNA and suppresses normal immunologic responses. Clinical signs depend on the magnitude and the actual wavelength of exposure and, to a lesser extent, on individual factors: e.g. darker skin can provide some protection against sunburn. Acute effects of ultraviolet radiation ranges from skin tan, red rash (erythema) to the various levels of sunburn. Chronic exposure can cause increased pigmentation, premature aging of the skin (photoaging), development of rough, scaly patches (solar keratosis) and squamous cell cancer, and less frequently basal cell carcinoma (basalioma). Skin cancers may be preceded by keratosis and keratoacanthoma. It is not thoroughly verified whether occupational exposures has anything to do with the development of melanoma: the findings are contradictory. Subjects of ultraviolet radiation are workers in open air and those exposed to artificial ultraviolet radiation (welding, UV lamps and lasers used in medicine, industry, trade and at home). Prevention of occupational skin cancers due to radiation consists of proper occupational safety and health organisation and instructions, shields, hats and (UV-absorbent) clothing together with medical surveillance. During medical check-ups keratoses can be discovered for treatment, while workers with verified skin cancers should be removed from exposure.

Occupational Burnout

Job burnout is a special type of job stress — a state of physical, emotional or mental exhaustion combined with doubts about your competence and the value of your work. If you think you might be experiencing job burnout, take a closer look at the phenomenon. What you learn might help you face the problem and take action before job burnout affects your health.

Ask yourself the following questions:

- Have you become cynical or critical at work?
- Do you drag yourself to work and have trouble getting started once you arrive?
- Have you become irritable or impatient with co-workers, customers or clients?
- Do you lack the energy to be consistently productive?
- Do you lack satisfaction from your achievements?
- Do you feel disillusioned about your job?
- Are you using food, drugs or alcohol to feel better or to simply not feel?
- Have your sleep habits or appetite changed?
- Are you troubled by unexplained headaches, backaches or other physical complaints?

If you answered yes to any of these questions, you might be experiencing job burnout. Be sure to consult your doctor or a mental health provider, however. Some of these symptoms can also indicate certain health conditions, such as a thyroid disorder or depression.

Causes of Job Burnout

Job burnout can result from various factors, including:

- Lack of control: An inability to influence decisions that affect your job — such as your schedule, assignments or workload — could lead to job burnout. So could a lack of the resources you need to do your work.
- Unclear job expectations: If you›re unclear about the degree of authority you have or what your supervisor or others expect from you, you›re not likely to feel comfortable at work.
- Dysfunctional workplace dynamics: Perhaps you work with an office bully, or you feel undermined by colleagues or your boss micromanages your work. This can contribute to job stress.
- Mismatch in values: If your values differ from the way your employer does business or handles grievances, the mismatch can eventually take a toll.

- Poor job fit. If your job doesn›t fit your interests and skills, it might become increasingly stressful over time.
- Extremes of activity: When a job is monotonous or chaotic, you need constant energy to remain focused — which can lead to fatigue and job burnout.
- Lack of social support: If you feel isolated at work and in your personal life, you might feel more stressed.
- Work-life imbalance: If your work takes up so much of your time and effort that you don't have the energy to spend time with your family and friends, you might burn out quickly.

Indications for Job Burnout

You might be more likely to experience job burnout if:

- You identify so strongly with work that you lack a reasonable balance between your work life and your personal life.
- You try to be everything to everyone.
- You work in a helping profession, such as health care, counseling or teaching.
- You feel you have little or no control over your work.
- Your job is monotonous.

Consequences of Job Burnout

Ignored or unaddressed job burnout can have significant consequences, including:

- Excessive stress
- Fatigue
- Insomnia
- A negative spillover into personal relationships or home life
- Depression
- Anxiety
- Alcohol or substance abuse
- Heart disease
- High cholesterol
- Type 2 diabetes, especially in women
- Stroke

- Obesity
- Vulnerability to illnesses

Remember, if you think you might be experiencing job burnout, don't ignore your symptoms. Consult your doctor or a mental health provider to identify or rule out any underlying health conditions.

Ways To Handle Job Burnout

If you're concerned about job burnout, take action. To get started:

- Manage the stressors that contribute to job burnout: Once you›ve identified what's fueling your feelings of job burnout, you can make a plan to address the issues.
- Evaluate your options: Discuss specific concerns with your supervisor. Perhaps you can work together to change expectations or reach compromises or solutions. Is job sharing an option? What about telecommuting or flexing your time? Would it help to establish a mentoring relationship? What are the options for continuing education or professional development?
- Adjust your attitude: If you›ve become cynical at work, consider ways to improve your outlook. Rediscover enjoyable aspects of your work. Recognize co-workers for valuable contributions or a job well-done. Take short breaks throughout the day. Spend time away from work doing things you enjoy.
- Seek support: Whether you reach out to co-workers, friends or loved ones, support and collaboration might help you cope with job stress and feelings of burnout. If you have access to an employee assistance program (EAP), take advantage of the available services.
- Assess your interests, skills and passions: An honest assessment can help you decide whether you should consider an alternative job, such as one that›s less demanding or one that better matches your interests or core values.
- Get some exercise: Regular physical activity can help you to better deal with stress. It can also help you get your mind off work and focus on something else.
- Get some sleep: Sleeps restores well-being and helps protect your health. Aim for at least 7-8 hours each night.

Occupational Asthma

Occupational asthma is asthma that's caused by breathing in chemical fumes, gases, dust or other substances on the job. Occupational asthma can result from exposure to a substance you're sensitive to — causing an allergic or immunological response — or to an irritating toxic substance.

Like other types of asthma, occupational asthma can cause chest tightness, wheezing and shortness of breath. People with allergies or with a family history of allergies are more likely to develop occupational asthma.

Avoidance of occupational triggers is an important part of management. Otherwise, treatment for occupational asthma is similar to treatment for other types of asthma and generally includes taking medications to reduce symptoms. If you already have asthma, sometimes treatment can help it from becoming worse in the workplace.

If it's not correctly diagnosed and you are not protected or able to avoid exposure, occupational asthma can cause permanent lung damage, disability or death.

Symptoms

Asthma Attack

Occupational asthma symptoms are similar to those caused by other types of asthma. Signs and symptoms may include:

- Wheezing, sometimes just at night
- Coughing
- Shortness of breath
- Chest tightness

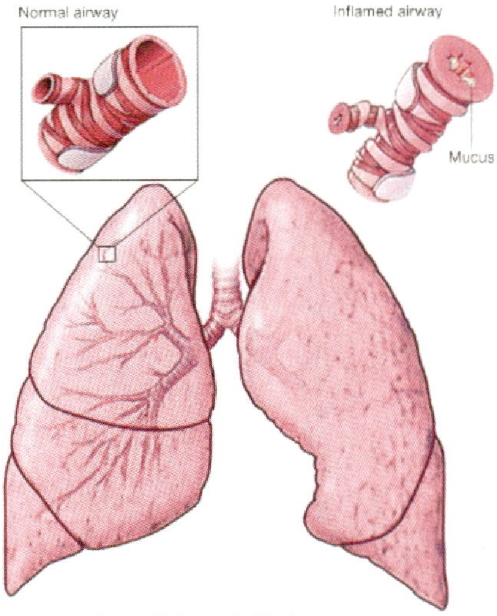

Other possible accompanying signs and symptoms may include:

- Runny nose

- Nasal congestion
- Eye irritation and tearing

Occupational asthma symptoms depend on the substance you're exposed to, how long and how often you're exposed, and other factors. Your symptoms may:

- Get worse as the workweek progresses, go away during weekends and vacations, and recur when you return to work.
- Occur both at work and away from work.
- Start as soon as you're exposed to an asthma-inducing substance at work or only after a period of regular exposure to the substance.
- Continue after exposure is stopped. The longer you're exposed to the asthma-causing substance, the more likely you'll have long-lasting or permanent asthma symptoms.

Jobs with the highest rates of occupational asthma:

- Vehicle spray painting - workers spraying cars and other vehicles are most often affected by a chemical called isocyanate (typically in 2-pack paints).
- Baking - workers in bakeries, flour mills and kitchens can be affected by flour dust and enzymes which contain additives such as amylase.
- Woodworking - people working in joinery and furniture industries can come into contact with dusts from hardwood, softwood and wood composite when it is machined or sanded.
- Soldering - people who work in the electronics and assembly industries can come into contact with fumes from rosin-based solder flux during their work.
- Healthcare work - there are two key substances that can cause asthma in healthcare workers. (1) Latex - natural rubber latex proteins have the potential to cause asthma if someone comes into contact with them regularly. (2) Diathermy - this surgical technique uses heat from an electric current to cut tissue or seal bleeding vessels. This process can give off invisible toxic gases, particles and vapours that may cause asthma symptoms.
- Working with animals - people who work with animals - from pet shops to stable owners, laboratory workers to zookeepers - can be affected by animal fur, feathers, dander, dried urine and saliva dusts during animal handling and cage or enclosure cleaning. These dusts contain proteins known as ‹animal aeroallergens› that can cause occupational asthma.
- Working in agriculture - agricultural dusts are the most common cause of occupational asthma in this industry. People often come into contact with grain dust

and poultry dust, or other dusts that are a mix of materials from fungal spores, bacteria, endotoxins, mites, animal dander and faeces, plant dust, soil, bedding, feed and feed components, chemicals, etc.

- Engineering - people who work in machining or shaping can be exposed to metal working fluids (MWFs) which can cause occupational asthma if you inhale the mist or vapour.
- Hairdressing - Hairdressers› bleach contains persulphate which can cause occupational asthma.

Diagnosis

To diagnose occupational asthma it is necessary to confirm the symptoms of asthma and establish the causal connection with the work environment. Various diagnostic tests can be used to aid in diagnoses of work related asthma.

A spirometer is a device used to measure timed expired and inspired volumes, and can be used to help diagnose asthma.

Peak expiratory flow rate (PEFR) is a hand held device which measures how fast a person can exhale and is a reliable test for occupational asthma. Serial PEFR can be measured to see if there is a difference in ability to exhale at work compared to that in a controlled environment.

A non-specific bronchial hyperreactivity test can be used to support the diagnose occupational asthma. It involves measuring the forced expiratory volume in 1 second (FEV-1) of the patient before and after exposure to methacholine or mannitol. Presence of airway responsiveness i.e. significant drop in FEV-1 can be seen in patients with occupational asthma.

Specific inhalation challenges test consist of exposing the subjects to the suspected occupational agent in the laboratory and/or at the workplace and assess for asthma symptoms as well as a reduction in FEV_1.

Other tests such as skin prick test, serum immunologic testing and measurement of sputum eosinophils can also be useful in establishing the diagnosis of occupational asthma.

Prevention and Management

Several forms of preventive measures have been suggested to prevent development of occupational asthma and also detect risk or disease early to allow intervention and improve outcomes. These include: comprehensive programs, education and training, medical examinations, use of medications, reduction of exposures and elimination of exposures.

Asthma symptoms and airway hyperresponsiveness can persist for several years after removal from the offending environment. Thus, early restriction from exposure to the trigger is advisable. Completely stopping exposure is more effective treatment than reducing exposure, but not always feasible.

Medication

Medications used for occupational asthma are similar to those used for other types of asthma such as short-acting beta-agonists like salbutamol or terbutaline, long-acting beta-agonists like salmeterol and formoterol and inhaled corticosteroids. Immunotherapy can also be used in some cases of sensitizer induced occupational asthma.

When a person is diagnosed with occupational asthma, it can result in serious socio-economic consequences not only for the workers but also for the employer and the healthcare system because the worker must change positions. The probability of being re-employed is lower for those with occupational asthma compared to those with normal asthma. The employer not only pays compensation to the employee, but will also have to spend a considerable amount of time and energy and funds for hiring and training new personnel. In the United States, it was estimated that the direct cost of occupational asthma in 1996 was $1.2 billion and the indirect cost $0.4 billion, for a total cost of $1.6 billion.

However, this can entail severe socio-economic consequences for the worker as well as the employer due to loss of job, unemployment, compensation issues, medical expenditures, and hiring and re-training of new personnel.

Prevention through Design

A way of minimizing workplace safety risks is the introduction of safety gear and personal protective equipment. This is however considered the least effective method of hazard control. An innovative way of ensuring good employee safety is the adoption of prevention by design or safety by design. It is a framework of methods that aims to minimize occupational risks early in the design developmental process. The implementation of these methods has to occur throughout the lifecycle of material resources and processes. An understanding of the risks to personal safety and an insight into healthy working conditions is of the utmost essence in the planning of safety measures. It supports the view that along with the considerations of cost and quality, safety has to be an essential feature in the design stage.

Each year in the U.S., 55,000 people die from work-related injuries and diseases, 294,000 are made sick, and 3.8 million are injured. The annual direct and indirect costs have been estimated to range from $128 billion to $155 billion. Recent studies in Australia indicate that design is a significant contributor in 37% of work-related fatalities; therefore, the successful implementation of prevention through design concepts can have substantial impacts on worker health and safety.

The National Institute for Occupational Safety and Health (NIOSH) in the United States is a major contributor and promoter of PtD policy and guidelines. NIOSH considers PtD to be "the most effective and reliable type" of prevention of occupational injuries. A core tenet of PtD philosophy the concept of addressing workplace hazards using methods at the top of the Hierarch of Controls, namely elimination and substitution.

Within Europe, construction designers are legally bound to design out risks during design development to reduce hazards in the construction and end use phases via the Mobile Worksite Directive (also known as CDM regulations in the UK). The concept supports this legal requirement. Some Notified Bodies provide testing and design verification services to ensure compliance with the safety standards defined in regulation codes such as the American Society of Mechanical Engineers. Many non-governmental organizations have been established to support this aim, principally in the UK, Australia and the United States.

While engineering as a rule factors human safety into the design process, a modern appraisal of specific links to design and workers' safety can be seen in efforts beginning in the 1800s. Trends included the widespread implementation of guards for machinery, controls for elevators, and boiler safety practices. This was followed by enhanced design for ventilation, enclosures, system monitors, lockout/tagout controls, and hearing protectors. More recently, there has been the development of chemical process safety, ergonomically engineered tools, chairs, and work stations, lifting devices, retractable needles, latex-free gloves, and a parade of other safety devices and processes.

In 2007, the National Institute for Occupational Health and Safety began its National Initiative on Prevention through Design with the goal of promoting prevention through design philosophy, practice, and policy.

Prevention through design represents a shift in approach for on-the-job safety. It involves evaluating potential risks associated with processes, structures, equipment, and tools. It takes into consideration the construction, maintenance, decommissioning, and disposal or recycling of waste material.

The idea of redesigning job tasks and work environments has begun to gain momentum in business and government as a cost-effective means to enhance occupational safety and health. Many U.S. companies openly support PtD concepts and have developed management practices to implement them. Other countries are actively promoting PtD concepts as well. The United Kingdom began requiring construction companies, project owners, and architects to address safety and health during the design phase of projects in 1994. Australia developed the Australian National OHS Strategy 2002–2012, which set "eliminating hazards at the design stage" as one of five national priorities. As a result, the

Australian Safety and Compensation Council (ASCC) developed the Safe Design National Strategy and Action Plans for Australia encompassing a wide range of design areas.

The National Institute for Occupational Safety and Health is a large contributor to prevention through design efforts in the United States. Several NIOSH initiatives and guidelines directly or indirectly advocate for PtD practices. Through NIOSH efforts, the U.S. Green Building Council posted new PtD credits available for Leadership in Energy and Environmental Design (LEED) certification for construction. Additionally, they provide a wide variety of educational and guidance materials on the topic of PtD The NIOSH "Buy Quiet" initiative uses elements of prevention through design to encourage companies to buy quieter machinery, thereby reducing occupational hearing loss for their workers.

Industrial Hygiene

Industrial hygiene is the science of protecting and enhancing the health and safety of people at work and in their communities. Health and safety hazards cover a wide range of chemical, physical, biological and ergonomic stressors. Those dedicated to anticipating, recognizing, evaluating and controlling those hazards are known as Industrial Hygienists. They are professionals dedicated to the well-being of people – at work, at home and in the community.

The term "industrial hygiene" gained popularity in the early 1900's with the formation of the Industrial Hygiene Section of the American Public Health Association. Since there were few "industrial hygienists", organizations began to use their Scientists (e.g. chemists, toxicologists, physicians, engineers) to evaluate and improve the health of the work environment, largely with a focus on chemical and noise hazards. The term industrial hygiene is still in use although its application and scope has expanded greatly since that time period to all hazards, industries, workers, their families and the communities. Today, there are other terms that essentially mean the same thing as industrial hygiene, e.g., Occupational Hygiene, Occupational Health, Environment, Health & Safety (EHS).

Anticipation, recognition, evaluation and control of hazards have always been the prime goal for the industrial hygiene professional. Traditionally, the Industrial Hygienist became proficient by some combination of education, training, and experience. Ideally, this knowledge is used to anticipate when a hazardous condition may occur and then presumably, take corrective action. But not everything can be anticipated, so the Industrial Hygienist must be able to recognize conditions that could lead to adverse health effects to workers or a community population. Once anticipated or recognized, the hazard could then be evaluated and eventually controlled to eliminate or reduce the impact of the hazard.

The Social Role of Occupational Hygiene

Occupational hygienists have been involved historically with changing the perception of society about the nature and extent of hazards and preventing exposures in the workplace and communities. Many occupational hygienists work day-to-day with industrial situations that require control or improvement to the workplace situation however larger social issues affecting whole industries have occurred in the past e.g. since 1900, asbestos exposures that have affected the lives of tens of thousands of people. Occupational hygienists have become more engaged in understanding and managing exposure risks to consumers from products with new regulations such as REACh (Registration, Evaluation, Authorisation and Restriction of Chemicals).

More recent issues affecting broader society are, for example in 1976, Legionnaires 'disease or legionellosis. More recently again in the 1990s radon and in the 2000s the effects of mold from indoor air quality situations in the home and at work. In the later part of the 2000s concern has been raised about the health effects of nanoparticles.

Many of these issues have required the coordination over a number of years of a number of medical and Para professionals in detecting and then characterizing the nature of the issue, both in terms of the hazard and in terms of the risk to the workplace and ultimately to society. This has involved occupational hygienists in research, collection of data and to develop suitable and satisfactory control methodologies.

General Activities

The occupational hygienist may be involved with the assessment and control of physical, chemical, biological or environmental hazards in the workplace or community that could cause injury or disease. Physical hazards may include noise, temperature extremes, illumination extremes, ionizing or non-ionizing radiation, and ergonomics. Chemical hazards related to dangerous goods or hazardous substances are frequently investigated by occupational hygienists. Other related areas including indoor air quality (IAQ) and safety may also receive the attention of the occupational hygienist. Biological hazards may stem from the potential for legionella exposure at work or the investigation of biological injury or effects at work, such as dermatitis may be investigated.

As part of the investigation process, the occupational hygienist may be called upon to communicate effectively regarding the nature of the hazard, the potential for risk, and the appropriate methods of control. Appropriate controls are selected from the hierarchy of control: by elimination, substitution, engineering, administration and personal protective equipment (PPE) to control the hazard or eliminate the risk. Such controls may involve recommendations as simple as appropriate PPE such as a 'basic' particulate dust mask to occasionally designing dust extraction ventilation systems, work places or management systems to manage people and programs for the preservation of health and well-being of those who enter a workplace.

Examples of occupational hygiene include:

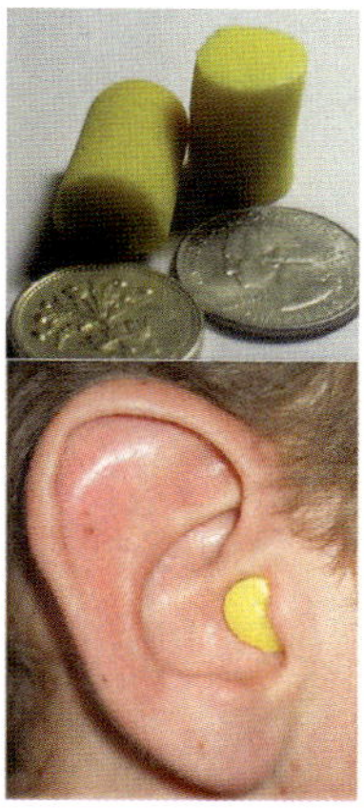

- Disposable foam earplugs: out of the ear with coins for scale (top) and inserted into the wearer's ear (bottom).
- Analysis of physical hazards such as noise that would lead to use of earplugs or earmuffs.
- Planning procedures to protect against infectious disease exposure in the event of a flu pandemic.
- Monitoring and testing the air for hazardous contaminants that can lead to potential worker illness and/or death.

Workplace Assessment Methods

Although there are many aspects to occupational hygiene work the most known and sought after is in determining or estimating potential or actual exposures to hazards. For many chemicals and physical hazards, occupational exposure limits have been derived using toxicological, epidemiological and medical data allowing hygienists to reduce the risks of health effects by implementing the "Hierarchy of Hazard Controls". Several methods can be applied in assessing the workplace or environment for exposure to a known or suspected hazard. Occupational hygienists do not rely on the accuracy of the equipment or method used but in knowing with certainty and precision the limits of the equipment or method being used and the error or variance given by using that particular equipment or method. Well known methods for performing occupational exposure assessments can be found in "A Strategy for Assessing and Managing Occupational Exposures, Third Edition Edited by Joselito S. Ignacio and William H. Bullock".

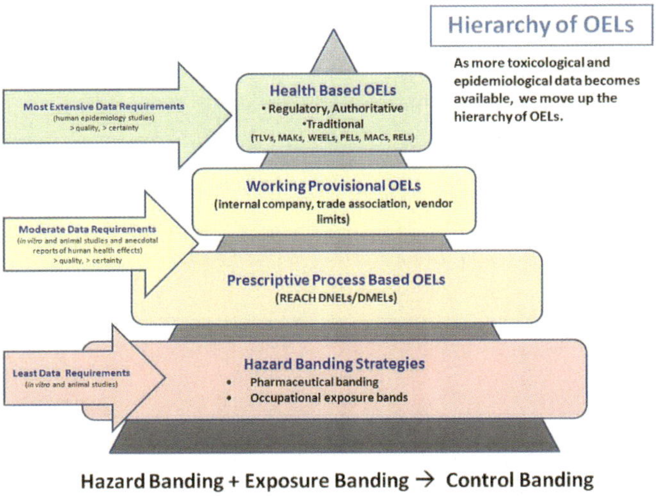

Hierarchy of occupational exposure limits (OELs)

The main steps outlined for assessing and managing occupational exposures:

- Basic Characterization (identify agents, hazards, people potentially exposed and existing exposure controls).
- Exposure Assessment (select occupational exposure limits, hazard bands, relevant toxicological data to determine if exposures are "acceptable", "unacceptable" or "uncertain").
- Exposure Controls (for "unacceptable" or "uncertain" exposures).
- Further Information Gathering (for "uncertain" exposures).
- Hazard Communication (for all exposures).
- Reassessment (as needed) / Management of Change.

Basic Characterization, Hazard Identification and Walk-through Surveys

The first step in understanding health risks related to exposures requires the collection of "basic characterization" information from available sources. A traditional method applied by occupational hygienists to initially survey a workplace or environment is used to determine both the types and possible exposures from hazards (e.g. noise, chemicals, radiation). The walk-through survey can be targeted or limited to particular hazards such as silica dust, or noise, to focus attention on control of all hazards to workers. A full walk-through survey is frequently used to provide information on establishing a framework for future investigations, prioritizing hazards, determining the requirements for measurement and establishing some immediate control of potential exposures. The Health Hazard Evaluation Program from the National Institute for Occupational Safety and Health is an example of an industrial hygiene walk-through survey. Other sources of basic characterization information include worker interviews, observing exposure tasks, material safety data sheets, workforce scheduling, production data, equipment and maintenance schedules to identify potential exposure agents and people possibly exposed.

The information that needs to be gathered from sources should apply to the specific type of work from which the hazards can come from. As mentioned previously, examples of these sources include interviews with people who have worked in the field of the hazard, history and analysis of past incidents, and official reports of work and the hazards encountered. Of these, the personnel interviews may be the most critical in identifying undocumented practices, events, releases, hazards and other relevant information. Once the information is gathered from a collection of sources, it is recommended for these to be digitally archived (to allow for quick searching) and to have a physical set of the same information in order for it to be more accessible. One innovative way to display the complex historical hazard information is with a historical hazards identification map, which distills the hazard information into an easy to use graphical format.

Sampling

Measurements of noise levels using a sound level meter is a component of the occupational hygiene assessment.

An occupational hygienist may use one or a number of commercially available electronic measuring devices to measure noise, vibration, ionizing and non-ionizing radiation, dust, solvents, gases, and so on. Each device is often specifically designed to measure a specific or particular type of contaminant. Such devices are often subject to multiple interferences. Electronic devices need to be calibrated before and after use to ensure the accuracy of the measurements taken and often require a system of certifying the precision of the instrument.

Dust Sampling

Nuisance dust is considered to be the total dust in air including inhalable and respirable fractions.

Various dust sampling methods exist that are internationally recognised. Inhalable dust is determined using the modern equivalent of the Institute of Occupational Medicine (IOM) MRE 113A monitor. Inhalable dust is considered to be dust of less than 100 micrometers aerodynamic equivalent diameter (AED) that enters through the nose and or mouth.

Respirable dust is sampled using a cyclone dust sampler design to sample for a specific fraction of dust AED at a set flow rate. The respirable dust fraction is dust that enters the 'deep lung' and is considered to be less than 10 micrometers AED.

Nuisance, inhalable and respirable dust fractions are all sampled using a constant volumetric pump for a specific sampling period. By knowing the mass of the sample collected and the volume of air sampled a concentration for the fraction sampled can be given in milligrams (mg) per metre cubed (m^3). From such samples the amount of inhalable or respirable dust can be determined and compared to the relevant occupational exposure limits.

By use of inhalable, respirable or other suitable sampler (7 hole, 5 hole, et cetera) these dust sampling methods can also used to determine metal exposure in the air. This requires collection of the sample on a methyl-cellulose ester (MCE) filter and acid digestion of the collection media in the laboratory followed by measuring metal concentration though an atomic absorption (or emission) spectrophotometry. Both the UK Health and Safety Laboratory and NIOSH Manual of Analytical Methods have specific methodologies for a broad range of metals in air found in industrial processing (smelting, foundries, et cetera).

A further method exists for the determination of asbestos, fibreglass, synthetic mineral fibre and ceramic mineral fibre dust in air. This is the membrane filter method (MFM) and requires the collection of the dust on a grided filter for estimation of exposure by the counting of 'conforming' fibres in 100 fields through a microscope. Results are quantified on the basis of number of fibres per millilitre of air (f/ml). Many countries strictly regulate the methodology applied to the MFM.

Chemical Sampling

Two types of chemically absorbent tubes are used to sample for a wide range of chemical substances. Traditionally a chemical absorbent 'tube' (a glass or stainless steel tube of between 2 and 10 mm internal diameter) filled with very fine absorbent silica (hydrophilic) or carbon, such as coconut charcoal (lipophilic), is used in a sampling line where air is drawn through the absorbent material for between four hours (minimum workplace sample) to 24 hours (environmental sample) period. The hydrophilic material readily absorbs water-soluble chemical and the lipophilic material absorbs non water-soluble materials. The absorbent material is then chemically or physically extracted and measurements performed using various gas chromatograph or mass spectrometry methods. These absorbent tube methods have the advantage of being usable for a wide range of potential contaminates. However, they are relatively expensive methods, are time consuming and require significant expertise in sampling and chemical analysis. A frequent complaint of workers is in having to wear the sampling pump (up to 1 kg) for several days of work to provide adequate data for the required statistical certainty determination of the exposure.

In the last few decades, advances have been made in 'passive' badge technology. These samplers can now be purchased to measure one chemical (e.g. formaldehyde) or a chemical type (e.g. ketones) or a broad spectrum of chemicals (e.g. solvents). They are relatively easy to set up and use. However, considerable cost can still be incurred in analysis of the 'badge'. They weigh 20 to 30 grams and workers do not complain about their presence. Unfortunately 'badges' may not exist for all types of workplace sampling that may be required and the charcoal or silica method may sometimes have to be applied.

From the sampling method, results are expressed in milligrams per cubic meter (mg/m^3) or parts per million (PPM) and compared to the relevant occupational exposure limits.

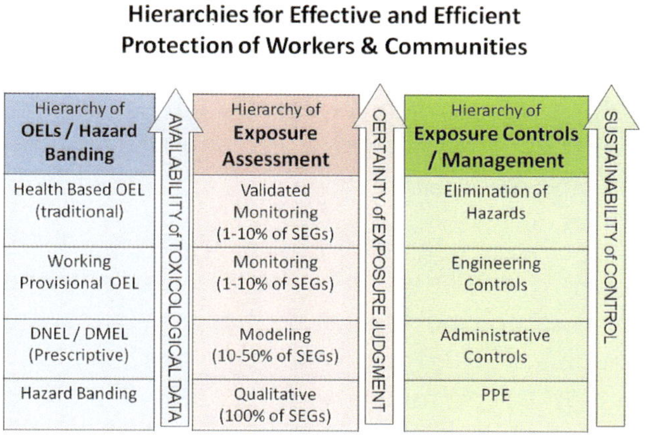

It is a critical part of the exposure determination that the method of sampling for the specific contaminate exposure is directly linked to the exposure standard used. Many countries regulate both the exposure standard, the method used to determine the exposure and the methods to be used for chemical or other analysis of the samples collected.

Simple representation of exposure risk assessment and management hierarchy based on available information.

Exposure Management and Controls

The hierarchy of control defines the approach used to reduce exposure risks protecting workers and communities. These methods include elimination, substitution, engineering controls (isolation or ventilation), administrative controls and personal protective equipment. Occupational hygienists, engineers, maintenance, management and employees should all be consulted for selecting and designing the most effective and efficient controls based on the hierarchy of control.

Defensible Space

A defensible space is an area around a building in which vegetation, debris, and other types of combustible fuels have been treated, cleared, or reduced to slow the spread of fire to and from the building. Information about local vegetation, weather, and topography is used to determine the Fire Severity Zone of an area, which can help determine the most effective design of a defensible space. A defensible space is one of the most cost-effective ways to protect a building from a wildfire and can often be created by the property owner.

Criteria

A first concept of Defensible Space for most fire agencies' primary goal of fuel reduction is a recommended or required defensible space around a structure to extend for at least 100 feet (30 m) in all directions.

- A second concept of Defensible Space is "Fuel Reduction." This means plants are selectively thinned and pruned to reduce the combustible fuel mass of the remaining plants. The goal is to break up the more continuous and dense uninterrupted layer of vegetation.
- A third concept of Defensible Space is "Fuel Ladder" management. Like rungs on a ladder, vegetation can be present at varying heights from groundcovers to trees. Ground fuel 'rungs' such as dried grasses, can transmit fire to shrub rungs, which then transmit up tree branch rungs into the tree canopy. A burning tree produces embers that can blow to new areas spreading and so making it more difficult to control a wildland fire. One guideline is for a typical separation of three times the height of the lower fuel to the next fuel ladder. For example, a 2-foot-high (0.61 m) shrub under a tree would need a spacing of 6 feet (1.8 m) to the lowest limbs of the tree. Since wildfires burn faster uphill than on flat land, fuel ladder spacing may need to be greater for slopes.

Landscape Use

The term Defensible Space in landscape ('firescape') use refers to the 100 feet (30 m) zone surrounding a structure. Often the location is in the wildland-urban interface. This area need not be devoid of vegetation by using naturally fire resistive plants that are spaced, pruned and trimmed, and irrigated, to minimize the fuel mass available to ignite and also to hamper the spread of a fire.

1. The first 30 feet (9.1 m) is the "Defensible Space Zone," of a defensible space around a structure. It is where vegetation is kept to a minimum combustible mass. A guideline used in this zone can be "low, lean and green."

2. The second distance of 30 to 100 feet (9.1 to 30.5 m), is the "Reduced Fuel Zone" of a defensible space around a structure. In this area of the defensible space, fuels/vegetation are separated vertically and horizontally depending on the vegetation type. This is done by: thinning, pruning, and removal of selected vegetation; and limbing up trees from lower vegetation and the lateral separation of tree canopies. types.

An important component is ongoing maintenance of the fire-resistant landscaping for reduced fuel loads and fire fighting access. Fire resistive plants that are not maintained can desiccate, die, or amass deadwood debris, and become fire assistive. Irrigation systems and pruning can help maintain a plant's fire resistance. Maintaining access roads and driveways clear of side and low-hanging vegetation can allow large fire equipment to reach properties and structures. Some agencies recommend clearing combustible vegetation at minimum horizontal 10 ft from roads and driveways a vertical of 13 ft 6 inches above them. Considering the plant material involved is important to not create unintended consequences to habitat integrity and unnecessary aesthetic issues. Street signs, and homes clearly identified with the numerical address, assist access also.

The slogan "Provide Defensible Space, be fire-safe and fire-smart" is used by agencies.

Unintended Consequences

The unintended negative consequences of erosion and native habitat loss can result from some unskillful defensible space applications. The disturbance of the soil surface, such as garden soil cultivation in and firebreaks beyond native landscape zones areas destroy the native plant cover and expose open soil, accelerating invasive species of plants ('invasive exotics') spreading and replacing native habitats.

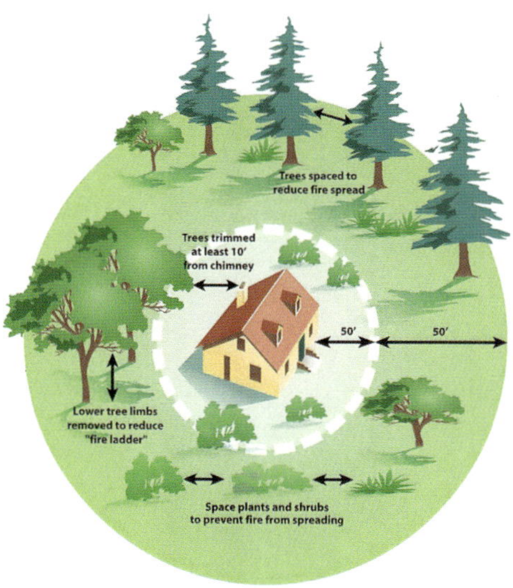

In suburban and wildland-urban interface areas the vegetation clearance and brush removal ordinances of municipalities for defensible space can result in mistaken excessive clear-cutting of native and non-invasive introduced shrubs and perennials that exposes the soil to more light and less competition for invasive plant species, and also erosion and landslides. Negative aesthetic consequences to natural and landscape areas can be minimized with integrated and balanced defensible space practices.

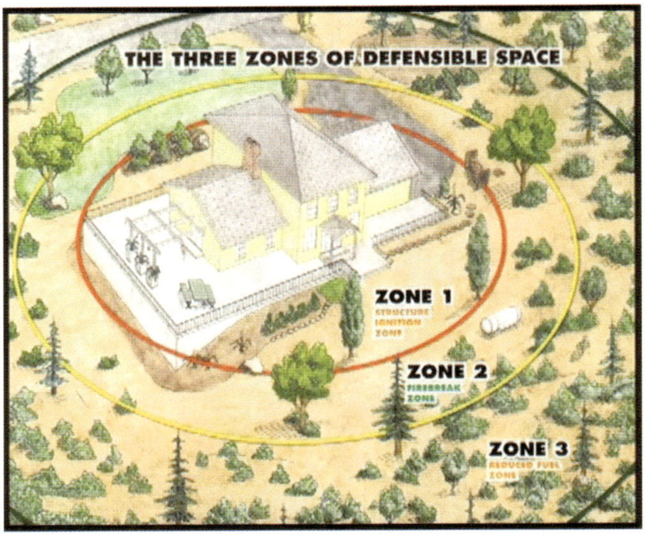

Engineering Controls

Engineering Control' can be described as a physical modification to a process, or process equipment, or the installation of further equipment with the goal of preventing the release of contaminants into the workplace (adapted from). As can be seen from this broad definition there are a wide range of engineering controls, which could be applied. The control selected will depend upon the type of process, the nature of the contaminant source (its toxicity and release mechanism) and the route of exposure (inhalation, dermal, and ingestion). However, the reality is that no single engineering control in isolation will be successful; control is always a mixture of equipment and ways of working.

The Hierarchy of Control

The approach to controlling the chemical risk released from a process is rarely straightforward as there will always be a choice of control options – some easier to apply than others. However, the approach taken should be based on a priority list. This principal of priority is often referred to as the 'Hierarchy of Control'. The

European Control Hierarchy, as stipulated by Council Directive 98/24/EC gives the priority order and is summarized below: 1. Elimination of hazardous substances 2. Substitution by a substance less hazardous 3. Design of appropriate work processes and engineering controls and use of adequate equipment and materials, so as to avoid or minimize the release of hazardous chemical agents which may present a risk to workers' safety and health at the place of work 4. Application of collective protection measures at the source of the risk, such as adequate ventilation and appropriate organizational measures 5. Where exposure cannot be prevented by other means, the application of individual protection measures including personal protective equipment (PPE).

The directive goes on to state that hazardous chemical agents shall be eliminated or reduced to a minimum by:

- The design and organization of systems of work at the workplace.
- The provision of suitable equipment for work with chemical agents and maintenance procedures which ensure the health and safety of workers at work.
- Reducing to a minimum the number of workers exposed or likely to be exposed.
- Reducing to a minimum the duration and intensity of exposure.
- Appropriate hygiene measures.
- Reducing the quantity of chemical agents present at the workplace to the minimum required for the type of work concerned.

It should be noted that this hierarchical approach is not unique to Europe and is adopted by safety professionals worldwide. From the above list it can be seen that engineering controls are integrated into steps 1 to 4. For example it can be argued that modifying a manufacturing process so as to eliminate the hazardous substance is a form of engineering control. However, it is common practice to associate engineering controls with steps 3 and 4: i.e. once elimination and substitution of chemical hazards have been considered. At times engineering controls may not offer adequate control and may need to be supplemented with other measures. Often this will take the form of PPE, which includes respiratory protection equipment (RPE). As can been seen from the priority list, PPE is the last step if all other interventions fail to offer sufficient protection. The problem with PPE is that it only protects the wearer. For RPE this is of particular concern as whilst the process operator may be protected from an airborne hazard, once it is released into the air it will inevitably pervade the workplace and therefore expose others who are likely to be unprotected. Furthermore for RPE to be effective it needs to be properly selected and correctly fitted, making training and user cooperation essential.

Types and Examples of Engineering Controls

Table: Examples of engineering controls

Process/Exposure source	Engineering control	Additional procedural control
Cleaning with solvent on rag	(i) Use a rag holder (ii) Provide a small bin with a lid for used rags.	(i) Check controls are used (ii) Safe disposal of waste
Dust spills from damaged sacks	Portable vacuum cleaners with HEPA filter	(i) Ensure vacuum is maintained and available for use (ii) Safe emptying of vacuum cleaner
Cutting-fluid mist from a lathe	Put an enclosure around the lathe and extract and filter the air and discharge to a safe place (*Protective gloves will also be required*)	(i) Train workers (e.g. It takes time for the mist to clear from the enclosures and this clearance time must be known) (ii) Check and maintain fluid quality (iii) Test and maintain controls (iv) Carry out health checks
Dust from disc cutter on stone worktop	(i) Carry out the process in an enclosure fitted with extraction, filter and extract to a safe place	(i) Test and maintain controls (ii) Train workers (iii) Carry out health checks
Transfer of volatile liquids	(i) Pumping rather than pouring (ii) Tight fitting lids to minimize evaporation	Regular checks and maintenance (e.g. Check for damage to lids seals)
Evaporation of liquid from an electroplating tank	A layer of plastic balls floating on the surface to reduce both evaporation and mists	Check and maintain controls

Non-ventilation Engineering Controls

Non-ventilation controls have the capability to reduce or eliminate process emission rate, for example the use of well-fitting lids to liquid containers. They can range from enclosures, seals, jigs and handling aids. However, engineering controls are frequently assumed to involve some form of ventilation control. This is unfortunate, as whilst ventilation controls can be effective and are the most commonly applied control to airborne contaminates, dismissing other non-ventilation engineering controls can be a costly mistake both in terms of financial and health cost.

There is probably no single reason why non-ventilation methods of control are overlooked, but perhaps the two main reasons are (i) the perceived need to alter the process and the potential ramifications this entails, whereas ventilation is often seen as something that can be retrofitted to any process, with little or no process modification. The reality is that this approach often leads to poor and erratic control. Designing a ventilation system to effectively control airborne contaminants requires specialist advice. All too often ventilation systems are badly designed and installed, poorly maintained and therefore frequently fails to provide adequate control; (ii) Ventilation is seen as a low cost option, which is not true. As well as the capital cost required to buy and install a ventilation system there are associated running costs. The latter includes power (required to drive the air mover plus that required to heat/cool air that is brought into the workplace to replace the air extracted) and the maintenance costs, such as replacement filter units.

Ventilation

Unlike non-ventilation control, ventilation is unlikely to affect the emission rate from a process; rather it is designed to control the contaminant once it has been released.

Ventilation controls are probably the most widely used method to control airborne contaminants in the workplace and can be divided in to two types: general ventilation and local exhaust ventilation.

General Ventilation

General ventilation is the introduction of clean air into the workplace that eventually replaces the contaminated air. General ventilation can be subdivided into two further types: dilution ventilation and displacement ventilation.

The aim of dilution ventilation is to uniformly mix the clean air that is continually introduced in to the workplace with the contaminated air in order to dilute the contaminant concentration to an acceptable level. Whilst this is an accepted form of engineering control, its application is limited to low toxicity sources that are usually diffused throughout the workplace and where the workers are a sufficient distance from the source.

Displacement ventilation is where air is introduced with the aim of replacing the contaminated air by clean air with little or no mixing. In practice this is difficult to achieve, particularly over large areas and therefore needs specialist assistance. Both of the above types of general ventilation tend to use a significant amount of air, which usually needs to be heated or cooled; consequently this type of ventilation is an expensive solution.

Local Exhaust Ventilation

Local Exhaust Ventilation is designed to capture, receive or contain the airborne contaminant at source before it has chance to enter the workers breathing zone or mix with the workplace air. As the control is applied as close to the source as possible, considerably less air is required when compared to general ventilation and consequently LEV is typically more effective in both terms of effectiveness and cost. For this reason LEV is preferred to general ventilation and should be considered to be higher in the hierarchy of controls.

Designing and Implementing Engineering Controls

Design Considerations

Whilst elimination or substitution of the chemical hazard is the most effective solution, it is recognized that this is not always possible or straightforward. Often the process relies on the chemical in question and therefore elimination or substitution will not be an option. Consequently engineering controls are applied that either:

- Reduce the emission rate from a process, usually by process modification (non-ventilation controls).
- Capture or containment of the emission once it has been released from the process, usually by enclosing and air extraction.

An example of reducing the contaminate emission rate is by the application of water to a stone cutting disk. This significantly reduces the emission of dust, creating instead liquid slurry. It should be noted in this example disposal of the liquid slurry may create a further hazard. An example of removal of the contaminant once it has been generated would be by the application of local exhaust ventilation. This would be achieved by enclosing the process as much as possible and extracting the airborne contaminant with a relatively low volume, high velocity extraction system. Interestingly these two forms of engineering control have similar effectiveness.

The examples above illustrate that there is always more than one engineering control approach that can be applied to any process and therefore it is important that the various engineering controls are collated and their suitability assessed before a solution is selected. When assessing the controls the following need to be considered:

- Effectiveness
- Reliability
- Ease of use
- Financial cost.

Estimating the effectiveness of a control measure is not always easy and can easily be misjudged. Figure below illustrates simply that as potential exposure risk increases, the control effectiveness must also increase. Failure to do so results in what is often referred to as the 'control gap' and it is this gap that results in worker exposure. The mismatch between the effectiveness of the engineering controls and the process risk can occur for a number of reasons, ranging from a lack of appreciation of the extent of the exposure risk from the process, to an over optimistic belief in the capability of the control measure.

Illustration of the consequence of mismatch between the risk of exposure and the engineering control effectiveness.

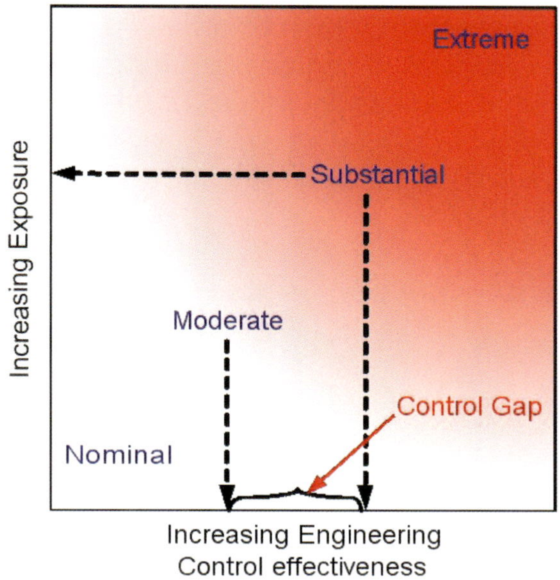

Adapted from HSE training material

Before engineering controls can be contemplated it is necessary to understand how the contaminant is being released into the workplace. This requires a full understanding of how the process works and how workers interact with the process. Ideally workers should be involved in the design and selection of the controls, as they will be using them on a daily basis. Failure to do this often results in engineering controls that are unworkable resulting in poor exposure control. Depending on the process, other disciplines may also need to be consulted, in particular the engineering function, as their input may be required to ensure the more intricate nuances of the process are considered and understood.

As well as considering the process, the reliability of any proposed control measures needs to be addressed. There is little point designing and installing what is judged to be an effective control solution that only works intermittently and is prone to malfunction.

Some engineering controls are more difficult and time consuming to apply than others and sometimes companies may have to resort to RPE equipment as a temporary control measure until the final engineering controls are design and implemented.

Additional Benefits of Good Control

Clearly engineering controls are designed to reduce exposure and to assist companies in complying with health and safety regulations and occupational exposure limits. However, it is possible that they may help to reduce environmental pollution and, importantly can make an economic impact by reducing company expenditure on such items as product consumables. An examples of this is given below.

There was a printing facility that introduced covers on older, high-solvent, printing machines. This engineering control required some thought, but hardly any capital expenditure. As a result the solvent vapour levels were halved, saving 5,000 litres of solvent per week equating to €74,400 a year. A clear case of an engineering control not only reducing worker exposure but saving the company a significant amount of money.

Ensuring that Engineering Controls are Effective and Reliable

Failures of engineering controls to protect their workers

Engineering controls can fail for a variety of reasons. Often they are not as effective as envisaged and therefore fail to protect from the date they are installed. Even when initially effective their performance can gradually decline. This can be exacerbated by poor management, e.g. inadequate training. Therefore there are issues to consider in ensuring controls work effectively and go on working.

Commissioning

Once a control measure is designed and installed it needs to be commissioned. 'Commissioning' is proving that the engineering control is capable of providing adequate exposure control. The type of commissioning and the complexity depends upon the control measure. Probably the most complex commissioning process is that of LEV systems. Unfortunately LEV commissioning is frequently carried out incompletely or is inadequate. LEV commissioning tends to focus on the engineering parameters, such as system pressures and air velocities. Whilst this is an essential part of the commissioning process, a judgment on the effectiveness of the controls and the worker exposure needs to be taken. There are a number of qualitative and quantitative tools available to help the assessor judge control. An example of a qualitative assessment is the use of smoke tubes to visualize the air flow in and around an LEV hood in order to assess LEV performance. An example of a quantitative control is personal sampling to quantify worker exposure to a particular substance(s).

Worker Training

In isolation, an engineering control solution is destined to fail. They need to be integrated with other control measures, such as a 'standard operating procedure'. It is highly likely that some form of training and supervision will also be required to ensure that the controls are correctly used and therefore control workers' exposure.

Checks, Monitoring and Maintenance

Without regular checks and routine maintenance, the effectiveness of engineering controls will degrade gradually and inevitably fail. The time it takes for this to occur will depend upon the type of control measure. Engineering controls tend to degrade slowly with time and this often goes unnoticed. An example of this are poorly maintained LEV systems; often the workers can hear the fan impeller rotating, but do not realize that the volume flow rate of the system is imperceptibly falling with time resulting in a loss of control of the airborne contaminant. In this example the performance of the LEV hood could be continually monitored by the use of an air flow indicator, such as a pressure gauge.

References

- Nikolai Fedotovich Izmerov, Vladimir Fedorovich Kirillov (editors); et al. (2016). "Introduction". Occupational health [Гигиена труда] (in Russian) (2 ed.). Moscow (RF): Publishing house ГЭОТАР-Медиа. p. 13, 14. ISBN 978-5-9704-3691-2
- Accidents, occupational-health: tyosuojelu.fi, Retrieved 17 May 2018
- Ruotsalainen, Jani H.; Verbeek, Jos H.; Mariné, Albert; Serra, Consol (2015-04-07). "Preventing occupational stress in healthcare workers". The Cochrane Database of Systematic Reviews (4): CD002892. doi:10.1002/14651858.CD002892.pub5. ISSN 1469-493X. PMID 25847433
- 7-most-common-causes-of-workplace-accidents: safetypartnersltd.com, Retrieved 14 April 2018
- "Ten Years of Working Conditions in the European Union, 2005". European Foundation for the Improvement of Living and Working Conditions. Retrieved 2007-12-01
- Top-10-Most-Common-Workplace-Injuries-202877: arbill.com, Retrieved 24 July 2018
- Pun K.-F.; Yam R.C.M.; Lewis W.G. (2003). "Safety management system registration in the shipping industry". International Journal of Quality & Reliability Management. 20 (6): 704–721. doi:10.1108/02656710310482140
- Occupational-asthma, understanding-asthma: asthma.org.uk, Retrieved 19 June 2018
- Chaturvedi, Pradeep (2006-01-01). Challenges of Occupational Safety and Health: Thrust : Safety in Transportation. Concept Publishing Company. ISBN 9788180692840
- Occupational-stress, counseling-process, counseling-psychology: psychology.iresearchnet.com, Retrieved 10 March 2018
- Karasek, R. A. (1979). "Job demands, job decision latitude, and mental strain-implications for job redesign". Administrative Science Quarterly. 24 (2): 285–308. doi:10.2307/2392498
- "Report on the American Workforce". United States Department of Labor, Bureau of Labor Statistics. Retrieved 2007-12-01

5

Industrial Safety Devices

Industrial safety devices are designed in order to protect employees from hazardous work conditions. This chapter closely examines the varied industrial safety devices that can be employed, such as blowout preventer, flame arrester, safety valve, chemical protective clothing, insulation monitoring device, etc.

Blowout Preventer

A blowout Preventer is a set of specialized valves installed on the well head in stacks, which are used to control the flow of gases or liquids from the well during drilling and production operations. BOP system is mainly operated during the situations of a well kick.

When drilling a well, mud system is a primary barrier and a BOP system is a secondary barrier for well control. When the primary barrier fails, hydrocarbons may enter the wellbore and a kick will occur. To gain control on this situation the drilling crew will operate the BOP system to close.

BOP is of two types- Annular and RAM type which comes in different sizes and pressure ratings. This is installed on top of the well head in stacks and allows access to the formation. They are designed to close the bore hole completely when there is no string in the well, allow closing of the well with the string (Drill pipe, Tubing etc.) inside and few are designed to shear or cut the string. BOP is a major equipment for safety of the rig and personnel, so they are inspected regularly and pressure tested every 21 days as per standards. On jack up rigs BOP is mounted below the rotary while in floater they are installed on surface of sea bed.

Functions of BOP:

- Control the flow of well fluids.
- Allow pumping of fluids into well.
- Monitor the wellbore pressure and allow to bleed of excess pressure.
- Shut in the well, i.e., seal the annular space between drill pipe and casing.
- Completely seal the well bore and shear the string.

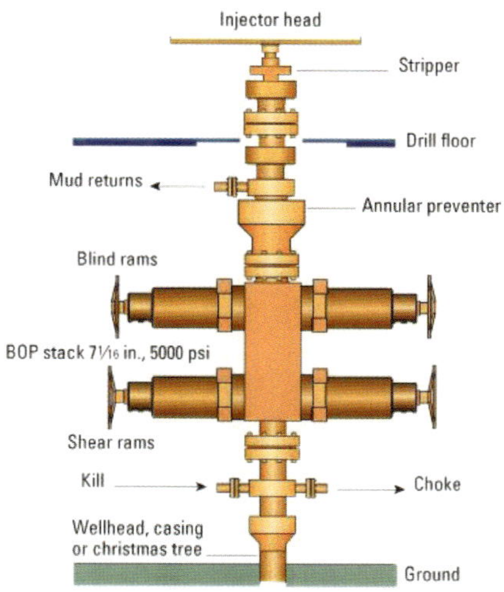

Additionally, and in performing those primary functions, blowout preventer systems are used to:

- Regulate and monitor wellbore pressure;
- Center and hang off the drill string in the wellbore;
- Shut in the well (e.g. seal the void, annulus, between drill pipe and casing);
- Kill the well (prevent the flow of formation fluid, influx, from the reservoir into the wellbore) ;
- Seal the wellhead (close off the wellbore);
- Sever the casing or drill pipe (in case of emergencies).

In drilling a typical high-pressure well, drill strings are routed through a blowout preventer stack toward the reservoir of oil and gas. As the well is drilled, drilling fluid, "mud", is fed through the drill string down to the drill bit, "blade", and returns up the wellbore in the ring-shaped void, annulus, between the outside of the drill pipe and the casing (piping that lines the wellbore). The column of drilling mud exerts downward hydrostatic pressure to counter opposing pressure from the formation being drilled, allowing drilling to proceed.

When a kick (influx of formation fluid) occurs, rig operators or automatic systems close the blowout preventer units, sealing the annulus to stop the flow of fluids out of the wellbore. Denser mud is then circulated into the wellbore down the drill string, up the annulus and out through the choke line at the base of the BOP stack through chokes (flow restrictors) until downhole pressure is overcome. Once "kill weight" mud extends from the bottom of the well to the top, the well has been "killed". If the integrity of the well is intact drilling may be resumed. Alternatively, if circulation is not feasible it may be possible to kill the well by "bullheading", forcibly pumping, in the heavier mud from the top through the kill line connection at the base of the stack. This is less desirable because of the higher surface pressures likely needed and the fact that much of the mud originally in the annulus must be forced into receptive formations in the open hole section beneath the deepest casing shoe.

If the blowout preventers and mud do not restrict the upward pressures of a kick, a blowout results, potentially shooting tubing, oil and gas up the wellbore, damaging the rig, and leaving well integrity in question.

Since BOPs are important for the safety of the crew and natural environment, as well as the drilling rig and the wellbore itself, authorities recommend, and regulations require, that BOPs be regularly inspected, tested and refurbished. Tests vary from daily test of functions on critical wells to monthly or less frequent testing on wells with low likelihood of control problems.

Exploitable reservoirs of oil and gas are increasingly rare and remote, leading to increased subsea deep water well exploration and requiring BOPs to remain submerged for as long as a year in extreme conditions. As a result, BOP assemblies have grown larger and heavier (e.g. a single ram-type BOP unit can weigh in excess of 30,000 pounds), while the space allotted for BOP stacks on existing offshore rigs has not grown commensurately. Thus a key focus in the technological development of BOPs over the last two decades has been limiting their footprint and weight while simultaneously increasing safe operating capacity.

Types

BOPs come in two basic types, ram and annular. Both are often used together in drilling rig BOP stacks, typically with at least one annular BOP capping a stack of several ram BOPs.

Ram Blowout Preventer

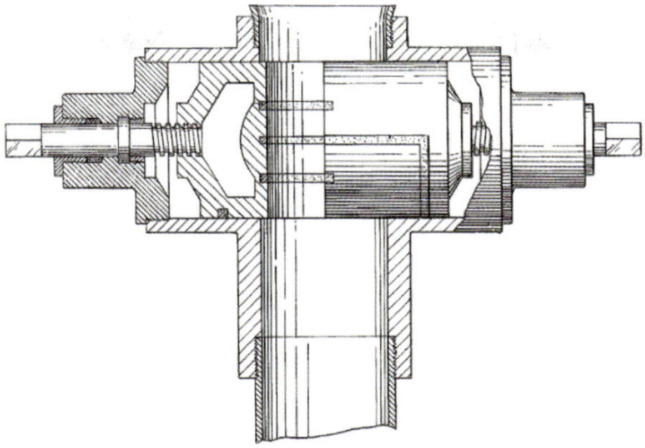

A Patent Drawing of the Original Ram-type Blowout Preventer, by Cameron Iron Works.

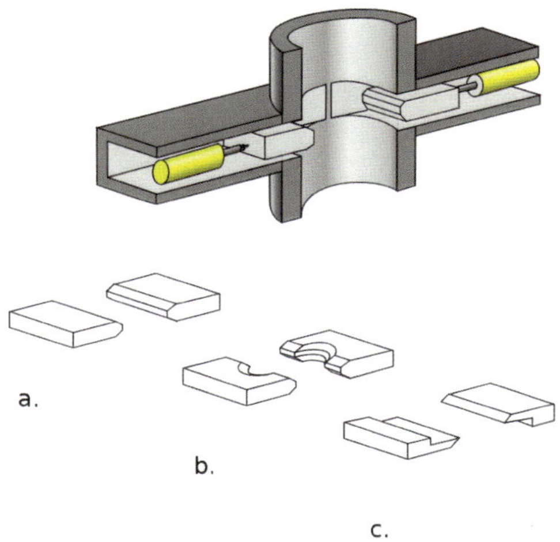

Blowout Preventer diagram showing different types of rams. (a) blind ram (b) pipe ram and (c) shear ram.

The ram BOP was invented by James Smither Abercrombie and Harry S. Cameron in 1922, and was brought to market in 1924 by Cameron Iron Works.

A ram-type BOP is similar in operation to a gate valve, but uses a pair of opposing steel plungers, rams. The rams extend toward the center of the wellbore to restrict flow or retract open in order to permit flow. The inner and top faces of the rams are fitted with packers (elastomeric seals) that press against each other, against the wellbore, and around tubing running through the wellbore. Outlets at the sides of the BOP housing (body) are used for connection to choke and kill lines or valves.

Rams, or ram blocks, are of four common types: pipe, blind, shear, and blind shear.

Pipe rams close around a drill pipe, restricting flow in the annulus (ring-shaped space between concentric objects) between the outside of the drill pipe and the wellbore, but do not obstruct flow within the drill pipe. Variable-bore pipe rams can accommodate tubing in a wider range of outside diameters than standard pipe rams, but typically with some loss of pressure capacity and longevity. Pipe ram should not be closed if there is no pipe in the hole.

Blind rams (also known as sealing rams), which have no openings for tubing, can close off the well when the well does not contain a drill string or other tubing, and seal it.

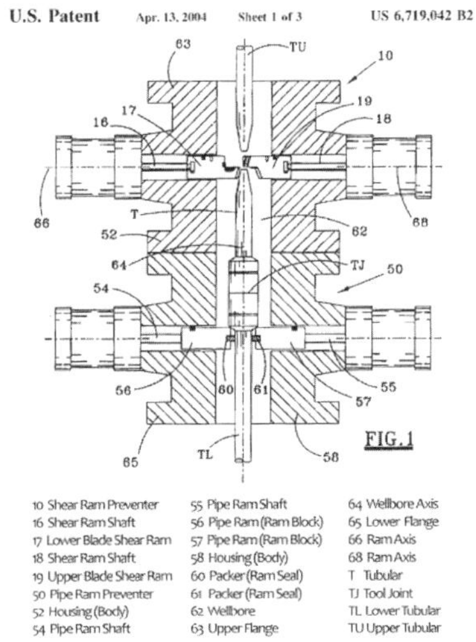

Patent Drawing of a Varco Shaffer Ram BOP Stack. A shear ram BOP has cut the drill-string and a pipe ram has hung it off.

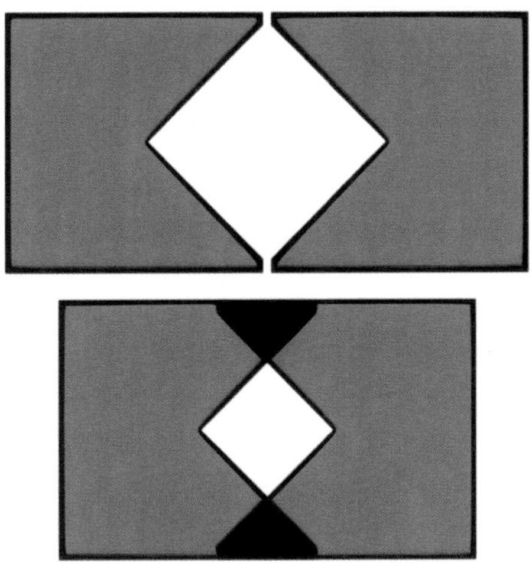

Schematic view of closing shear blades

Shear rams are designed to shear the pipe in the well and seal the wellbore simultaneously. It has steel blades to shear the pipe and seals to seal the annulus after shearing the pipe.

Blind shear rams (also known as shear seal rams, or sealing shear rams) are intended to seal a wellbore, even when the bore is occupied by a drill string, by cutting through the drill string as the rams close off the well. The upper portion of the severed drill string is freed from the ram, while the lower portion may be crimped and the "fish tail" captured to hang the drill string off the BOP.

In addition to the standard ram functions, variable-bore pipe rams are frequently used as test rams in a modified blowout preventer device known as a stack test valve. Stack test valves are positioned at the bottom of a BOP stack and resist downward pressure (unlike BOPs, which resist upward pressures). By closing the test ram and a BOP ram about the drill string and pressurizing the annulus, the BOP is pressure-tested for proper function.

The original ram BOPs of the 1920s were simple and rugged manual devices with minimal parts. The BOP housing (body) had a vertical well bore and horizontal ram cavity (ram guide chamber). Opposing rams (plungers) in the ram cavity translated horizontally, actuated by threaded ram shafts (piston rods) in the manner of a screw jack. Torque from turning the ram shafts by wrench or hand wheel was converted to linear motion and the rams, coupled to the inner ends of the ram shafts, opened and closed the well bore. Such screw jack type operation provided enough mechanical advantage for rams to overcome down hole pressures and seal the wellbore annulus.

Hydraulic rams BOPs were in use by the 1940s. Hydraulically actuated blowout preventers had many potential advantages. The pressure could be equalized in the opposing hydraulic cylinders causing the rams to operate in unison. Relatively rapid actuation and remote control were facilitated, and hydraulic rams were well-suited to high pressure wells.

Because BOPs are depended on for safety and reliability, efforts to minimize the complexity of the devices are still employed to ensure longevity. As a result, despite the ever-increasing demands placed on them, state of the art ram BOPs are conceptually the same as the first effective models, and resemble those units in many ways.

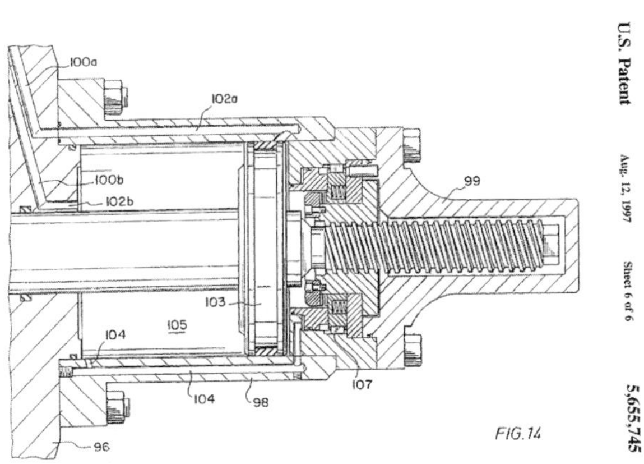

Hydril Company's Compact BOP Ram Actuator Assembly Patent Drawing

Ram BOPs for use in deepwater applications universally employ hydraulic actuation. Threaded shafts are often still incorporated into hydraulic ram BOPs as lock rods that hold the ram in position after hydraulic actuation. By using a mechanical ram locking mechanism, constant hydraulic pressure need not be maintained. Lock rods may be coupled to ram shafts or not, depending on manufacturer. Other types of ram locks, such as wedge locks, are also used.

Typical ram actuator assemblies (operator systems) are secured to the BOP housing by removable bonnets. Unbolting the bonnets from the housing allows BOP maintenance and facilitates the substitution of rams. In that way, for example, a pipe ram BOP can be converted to a blind shear ram BOP.

Shear-type ram BOPs require the greatest closing force in order to cut through tubing occupying the wellbore. Boosters (auxiliary hydraulic actuators) are frequently mounted to the outer ends of a BOP's hydraulic actuators to provide additional shearing force for shear rams.

Ram BOPs are typically designed so that well pressure will help maintain the rams in their closed, sealing position. That is achieved by allowing fluid to pass through a channel in the ram and exert pressure at the ram's rear and toward the center of the wellbore. Providing a channel in the ram also limits the thrust required to overcome well bore pressure.

Single ram and double ram BOPs are commonly available. The names refer to the quantity of ram cavities (equivalent to the effective quantity of valves) contained in the unit. A double ram BOP is more compact and lighter than a stack of two single ram BOPs while providing the same functionality, and is thus desirable in many applications. Triple ram BOPs are also manufactured, but not as common.

Technological development of ram BOPs has been directed towards deeper and higher pressure wells, greater reliability, reduced maintenance, facilitated replacement of components, facilitated ROV intervention, reduced hydraulic fluid consumption, and improved connectors, packers, seals, locks and rams. In addition, limiting BOP weight and footprint are significant concerns to account for the limitations of existing rigs.

The highest-capacity large-bore ram blowout preventer on the market, as of July 2010, Cameron's EVO 20K BOP, has a hold-pressure rating of 20,000 psi, ram force in excess of 1,000,000 pounds, and a well bore diameter of 18.75 inches.

Annular Blowout Preventer

Diagram of an Annular blowout preventer in open and fully closed configurations. The flexible annulus (donut) in blue is forced into the drill pipe cavity by the hydraulic pistons.

The annular blowout preventer was invented by Granville Sloan Knox in 1946; a U.S.

patent for it was awarded in 1952. Often around the rig it is called the "Hydril", after the name of one of the manufacturers of such devices.

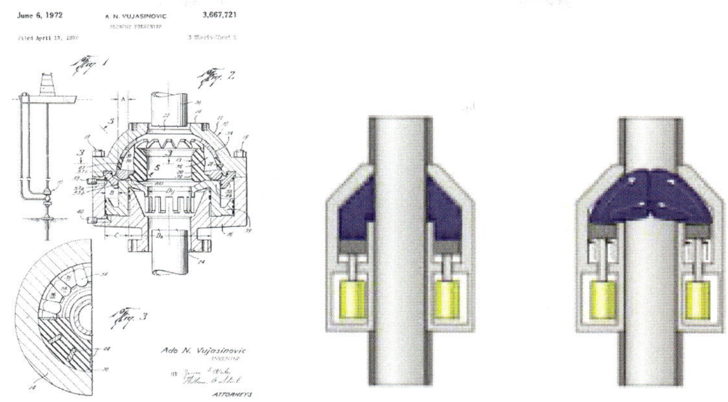

Patent Drawing of Original Shaffer Spherical-type Blowout Preventer

An annular-type blowout preventer can close around the drill string, casing or a non-cylindrical object, such as the kelly. Drill pipe including the larger-diameter tool joints (threaded connectors) can be "stripped" (i.e., moved vertically while pressure is contained below) through an annular preventer by careful control of the hydraulic closing pressure. Annular blowout preventers are also effective at maintaining a seal around the drillpipe even as it rotates during drilling. Regulations typically require that an annular preventer be able to completely close a wellbore, but annular preventers are generally not as effective as ram preventers in maintaining a seal on an open hole. Annular BOPs are typically located at the top of a BOP stack, with one or two annular preventers positioned above a series of several ram preventers.

An annular blowout preventer uses the principle of a wedge to shut in the wellbore. It has a donut-like rubber seal, known as an elastomeric packing unit, reinforced with steel ribs. The packing unit is situated in the BOP housing between the head and hydraulic piston. When the piston is actuated, its upward thrust forces the packing unit to constrict, like a sphincter, sealing the annulus or openhole. Annular preventers have only two moving parts, piston and packing unit, making them simple and easy to maintain relative to ram preventers.

The original type of annular blowout preventer uses a "wedge-faced" (conical-faced) piston. As the piston rises, vertical movement of the packing unit is restricted by the head and the sloped face of the piston squeezes the packing unit inward, toward the center of the wellbore.

In 1972, Ado N. Vujasinovic was awarded a patent for a variation on the annular preventer known as a spherical blowout preventer, so-named because of its spherical-faced head. As the piston rises the packing unit is thrust upward against the curved head, which constricts the packing unit inward. Both types of annular preventer are in common use.

Control Methods

When wells are drilled on land or in very shallow water where the wellhead is above the water line, BOPs are activated by hydraulic pressure from a remote accumulator. Several control stations will be mounted around the rig. They also can be closed manually by turning large wheel-like handles.

In deeper offshore operations with the wellhead just above the mudline on the sea floor, there are five primary ways by which a BOP can be controlled. The possible means are:

- Hydraulic Control Signal: sent from surface through a hydraulic umbilical.
- Electrical Control Signal: sent from the surface through a control cable.
- Acoustical Control Signal: sent from the surface based on a modulated/encoded pulse of sound transmitted by an underwater transducer.
- ROV Intervention: remotely operated vehicles (ROVs) mechanically control valves and provide hydraulic pressure to the stack (via "hot stab" panels).
- Deadman Switch / Auto Shear: fail-safe activation of selected BOPs during an emergency, and if the control, power and hydraulic lines have been severed.

Two control pods are provided on the BOP for redundancy. Electrical signal control of the pods is primary. Acoustical, ROV intervention and dead-man controls are secondary.

An emergency disconnect system/sequence, or EDS, disconnects the rig from the well in case of an emergency. The EDS is also intended to automatically trigger the deadman switch, which closes the BOP, kill and choke valves. The EDS may be a subsystem of the BOP stack's control pods or separate.

Pumps on the rig normally deliver pressure to the blowout preventer stack through hydraulic lines. Hydraulic accumulators are on the BOP stack enable closure of blowout preventers even if the BOP stack is disconnected from the rig. It is also possible to trigger the closing of BOPs automatically based on too high pressure or excessive flow.

Individual wells along the U.S. coastline may also be required to have BOPs with backup acoustic control. General requirements of other nations, including Brazil, were drawn to require this method. BOPs featuring this method may cost as much as US$500,000 more than those that omit the feature.

Deepwater Horizon Blowout

During the Deepwater Horizon drilling rig explosion incident on April 20, 2010, the blowout preventer should have been activated automatically, cutting the drillstring and sealing the well to preclude a blowout and subsequent oil spill in the Gulf of Mexico, but it failed to fully engage. Underwater robots (ROVs) later were used to manually trigger the blind shear ram preventer, to no avail.

A robotic arm of a Remotely Operated Vehicle (ROV) attempts to activate the "Deepwater Horizon" Blowout Preventer (BOP)

As of May 2010 it was unknown why the blowout preventer failed. Chief surveyor John David Forsyth of the American Bureau of Shipping testified in hearings before the Joint Investigation of the Minerals Management Service and the U.S. Coast Guard investigating the causes of the explosion that his agency last inspected the rig's blowout preventer in 2005. BP representatives suggested that the preventer could have suffered a hydraulic leak. Gamma-ray imaging of the preventer conducted on May 12 and May 13, 2010 showed that the preventer's internal valves were partially closed and were restricting the flow of oil. Whether the valves closed automatically during the explosion or were shut manually by remotely operated vehicle work is unknown. A statement released by Congressman Bart Stupak revealed that, among other issues, the emergency disconnect system (EDS) did not function as intended and may have malfunctioned due to the explosion on the Deepwater Horizon.

The permit for the Macondo Prospect by the Minerals Management Service in 2009 did not require redundant acoustic control means. Inasmuch as the BOPs could not be closed successfully by underwater manipulation (ROV Intervention), pending results of a complete investigation, it is uncertain whether this omission was a factor in the blowout.

Documents discussed during congressional hearings June 17, 2010, suggested that a battery in the device's control pod was flat and that the rig's owner, Transocean, may have "modified" Cameron's equipment for the Macondo site (including incorrectly routing hydraulic pressure to a stack test valve instead of a pipe ram BOP) which increased the risk of BOP failure, in spite of warnings from their contractor to that effect. Another hypothesis was that a junction in the drilling pipe may have been positioned in the BOP stack in such a way that its shear rams had an insurmountable thickness of material to cut through.

It was later discovered that a second piece of tubing got into the BOP stack at some point during the Macondo incident, potentially explaining the failure of the BOP shearing mechanism. As of July 2010 it was unknown whether the tubing might have been casing that shot up through the well or perhaps broken drill pipe that dropped into the well. The DNV final report indicated that the second tube was the segment of the drill string that was ejected after being cut by the blow out preventer shears.

On July 10, 2010 BP began operations to install a sealing cap, also known as a capping stack, atop the failed blowout preventer stack. Based on BP's video feeds of the operation the sealing cap assembly, called Top Hat 10, included a stack of three blind shear ram BOPs manufactured by Hydril (a GE Oil & Gas company), one of Cameron's chief competitors. By July 15 the 3 ram capping stack had sealed the Macondo well, if only temporarily, for the first time in 87 days.

The U.S. government wanted the failed blowout preventer to be replaced in case of any pressure change that occurs when the relief well intersected with the well. On September 3, 2010, at 1:20 p.m. CDT the 300 ton failed blowout preventer was removed from the well and began being slowly lifted to the surface. Later that day a replacement blowout preventer was placed on the well. On September 4 at 6:54 p.m. CDT the failed blowout preventer reached the surface of the water and at 9:16 p.m. CDT it was placed in a special container on board the vessel Helix Q4000. The failed blowout preventer was taken to a NASA facility in Louisiana for examination by Det Norske Veritas (DNV).

On 20 March 2011, DNV presented their report to the US Department of Energy. Their primary conclusion was that while the rams succeeded in partly shearing through the drill pipe they failed to seal the bore because the drill pipe had buckled out of the intended line of action of the rams (because the drill string was caught at a tool joint in the upper annular BOP valve), jamming the shears and leaving the drill string shear actuator unable to deliver enough force to complete its stroke and fold the cut pipe over and seal the well. They did not suggest any failure of actuation as would be caused by faulty batteries. The upper section of the blow out preventer failed to separate as designed due to numerous oil leaks compromising hydraulic actuator operation, and this had to be cut free during recovery.

Flame Arrester

A Flame Arrester also spelled arrestor is a device which allows gas to pass through it but stops a flame in order to prevent a larger fire or explosion. There is an enormous variety of situations in which flame arresters are applied. Anyone involved in selecting flame arresters needs to understand how these products work and their performance limitations.

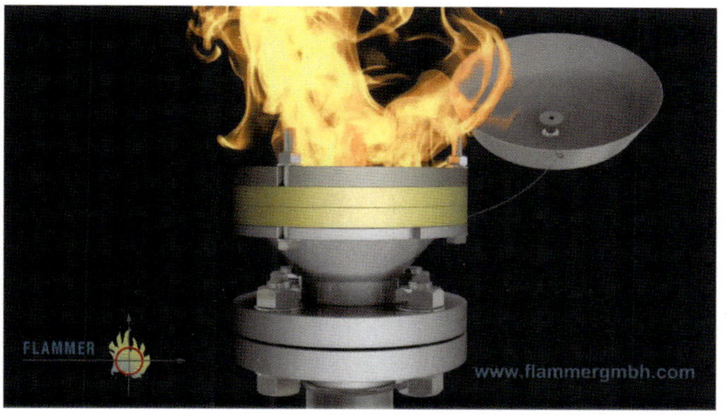

The operating principle of flame arresters was discovered in 1815 by Sir Humphry Davy, a famous chemist and professor at the Royal Institution in England. A safety committee of the English coal mining industry had approached Davy for technical assistance. They needed a way to prevent miners' oil lamps from causing explosions when flammable gas called firedamp seeped into the mine shafts. Sir Humphry studied the gas, which consisted mostly of methane. The investigation centered on how methane burns under various conditions and with various proportions of air. Davy's solution was to enclose the lamp flame securely with a tall cylinder of finely woven wire screen called metal gauze. Two of the earliest Davy safety lamps are shown on the right.

Enough lamplight passes out through the screen to be useful. Air for the oil flame around the lamp wick enters through the lower part of the screen. Hot exhaust gas escapes through the upper part. When a combustible mixture of methane flows in with the air, a methane flame burns against the inside of the screen. However, neither the methane flame nor the lamp flame passes through the narrow openings of the screen. The metal wire absorbs heat from the flame and then radiates it away at a much lower temperature.

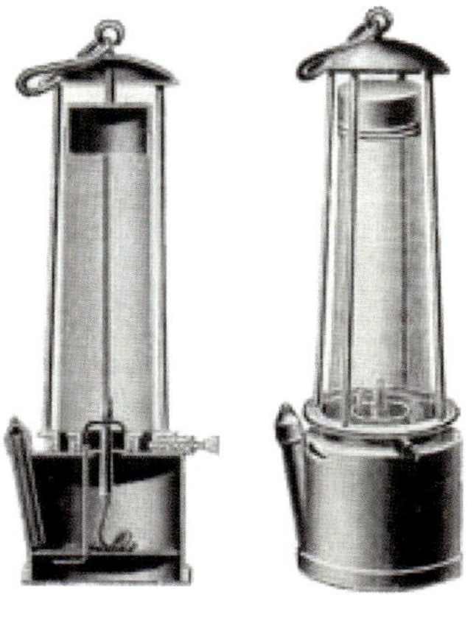

Modern Flame Arresters

Since Sir Humphry's time, flame arresters of numerous varieties have been applied in many industries. All of them operate on the same principle: removing heat from the flame as it attempts to travel through narrow passages with walls of metal or other heat-conductive material. For instance, flame arresters made by most manufacturers employ layers of metal ribbons with crimped corrugations.

Flame arresters are used in many industries, including refining, pharmaceutical, chemical, petrochemical, pulp and paper, oil exploration and production, sewage treatment, landfills, mining, power generation, and bulk liquids transportation. In some cases, the flames involve exothermic (heat-producing) reactions other than oxidation. Processes which generate the combustible or reactive gases include blending, reacting, separation, mixing, drilling, and digesting. These processes involve numerous equipment configurations and gas mixtures.

Working of Modern Flame Arresters

Flame arresters are passive devices with no moving parts. They prevent the propagation of flame from the exposed side of the unit to the protected side by the use of wound crimped metal ribbon type flame cell element.

This construction produces a matrix of uniform openings that are carefully constructed to quench the flame by absorbing the heat of the flame. This provides an extinguishing barrier to the ignited vapour mixture.

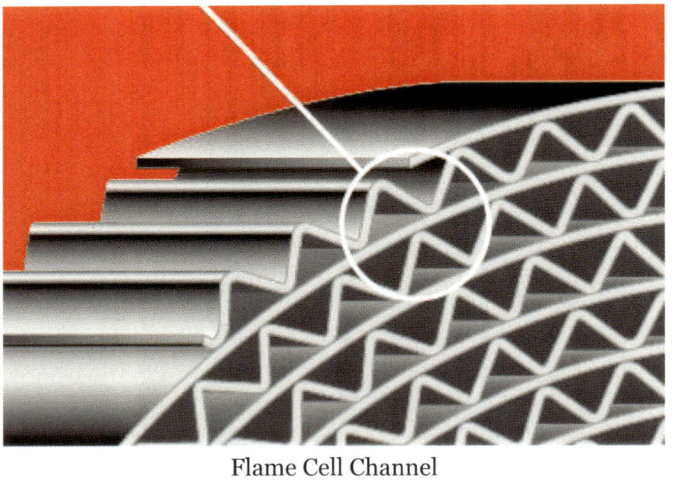

Flame Cell Channel

Under normal operating conditions the flame arrester permits a relatively free flow of gas or vapour through the piping system. If the mixture is ignited and the flame begins to travel back through the piping, the arrester will prohibit the flame from moving back to the gas source.

In-line Deflagration or Detonation Flame Arrester

The other major category consists of in-line flame arresters, also known as deflagration and detonation flame arresters. (Speaking non-technically, deflagration means rapid burning, and detonation means explosion.) These units are installed in pipes to prevent flames from passing.

Most in-line flame arrester applications are in systems which collect gases emitted by liquids and solids. These systems, commonly used in many industries, may be called vapor control systems. The gases which are vented to atmosphere or controlled via vapor control systems are typically flammable. If the conditions are such that ignition occurs, a flame inside or outside of the system could result, with the potential to do catastrophic damage.

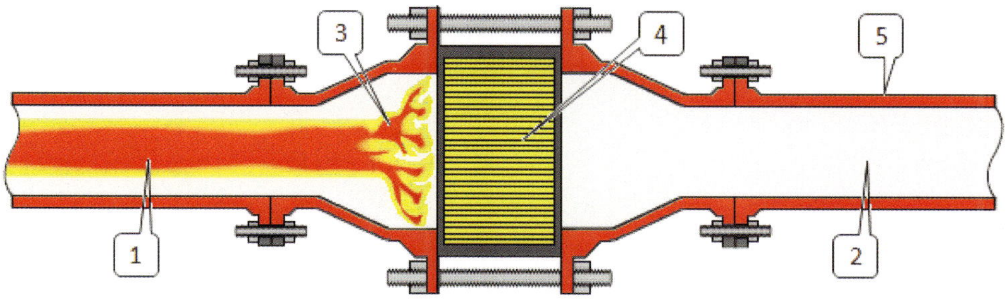

1. Exposed Side 2. Protected Side 3. Flame stabilized on arrester element 4. Flame arrester element absorbs and quenches flame front 5. Piping

One variety of vapor control systems is called vapor destruction systems. Included are elevated flare systems, enclosed flare systems, burner and catalytic incineration systems, and waste gas boilers.

Another type of vapor control system using in-line flame arresters is vapor recovery systems. Included here are vapor balancing, refrigeration, adsorption, absorption, and compression systems.

However, in-line flame arresters are sometimes used in end-of-line applications. For instance, an in-line unit may be mounted below a tank vent Valve on a liquid storage tank. The Valve reduces emissions and product loss, while the flame arrester protects the tank from flames in the atmosphere during venting of flammable gases.

Selecting In-Line Flame Arresters

The various dynamic states explained earlier for confined flames can be very dangerous for a process system due to the tremendous energies associated with detonation pressure and flame velocity. Things happen fast and can turn catastrophic. These multiple dynamic states increase the challenge of providing a flame arrester product or products which stop the flame and withstand the enormous pressures caused by explosions within the confined piping.

The very wide range of possible behavior for a confined flame causes two particular problems for flame arrester products. First, the high-pressure deflagration and stable detonation states have very stable kinetics of burning, and the flame is moving very fast. Therefore the arrester must be able to absorb the flame's heat much faster than is required by standard low-to-medium-pressure deflagration conditions. Second, the instantaneous impulse pressures caused by the shock waves of overdriven detonation subject the arrester to forces of up to 20995 kPa(g) (3000 psig). Thus, the arrester must be structurally superior to standard lowpressure deflagration arresters.

End of Line or Vent-to-Atmosphere Flame Arrester

End of line or vent-to-atmosphere flame arresters allow free venting in combination with flame protection for vertical vent applications. They prevent flame propagation by absorbing and dissipating heat using spiral wound crimped ribbon stainless steel flame cells. End-of-line flame arresters are used in applications such as petroleum storage tank vents.

The classic application is in preventing fire in the atmosphere from entering an enclosure. Around 1920, for instance, flame arresters began to be installed on vents on oilfield storage tanks. They keep the tanks from exploding when gas flowing from the vents is struck by lightning.

Conversely, some end-of-line flame arresters prevent fire in an enclosure from igniting an explosive atmosphere such as in a refinery. For instance, flame arresters may be installed in furnace air inlets and exhaust stacks.

Selecting End-of-Line Flame Arresters

End-of-line deflagration flame arresters are designed for unconfined flame propagation, also referred to as atmospheric explosion or unconfined deflagration. They simply bolt or screw onto the process or tank connection. These designs incorporate well-established but simple technology. Most use a single element of crimped wound metal ribbon that provides the Heat Transfer needed to quench the flame before it gets through the arrester element.

The main points of concern when selecting an arrester for end-of-line applications are as follows:

- Hazardous group designation or MESG value of the gas.
- Flame stabilization performance characteristics of the arrester compared to the system potential for flame stabilization for sustained periods of time.
- Process gas temperature.
- Pressure drop across the arrester during venting flow conditions, relative to the system's maximum allowable pressure and vacuum.
- Materials of construction that meet the ambient and process conditions - for example, extremely cold climate, salt spray, chemically aggressive gas, etc.
- Connection type and size.
- Instrumentation requirements.

Usage and Applications

Flame arresters are used:

- To stop the spread of an open fire.
- To limit the spread of an explosive event that has occurred.
- To protect potentially explosive mixtures from igniting.
- To confine fire within an enclosed, controlled, or regulated location.
- To stop the propagation of a flame traveling at sub-sonic velocities.

They are commonly used on:

- Fuel storage tank vents.
- Fuel gas pipelines.
- Safety storage cabinets for paint, aerosol cans, and other flammable mixtures.
- The exhaust system of internal combustion engines.
- The air intake of marine inboard engines.
- Davy lamps in coal mining.
- Overproof rum and other flammable liquors.

Principles

A flame arrester (also called a deflagration arrester) functions by absorbing the heat from a flame front traveling at sub-sonic velocities, thus dropping the burning gas/air mixture below its auto-ignition temperature: consequently, the flame cannot survive. The heat is absorbed through channels (passages) designed into an element. These channels are chosen and measured as the mesg (Maximum Experimental Safe Gap) of the gas for a particular installation. These passages can be regular, like crimped metal ribbon or wire mesh or a sheet metal plate with punched holes, or irregular, such as those in random packing.

The required size of the channels needed to stop the flame front can vary significantly, depending on the flammability of the fuel mixture. The large openings on a chain link fence are capable of slowing the spread of a small, slow-burning grass fire, but fast-burning grass fires will penetrate the fence unless the holes are very small. In a coal mine containing highly explosive coal dust or methane, the wire mesh of a Davy lamp must be very tightly spaced.

For flame arresters used as a safety device, the mesh must be protected from damage due to being dropped or struck by another object, and the mesh must be capable of rigidly retaining its shape during the propagation of a flame front. Any shifting of the individual wires that make up the mesh can create an opening large enough to allow the flame to penetrate and spread beyond the barrier.

On a fuel storage vent, flame arresters also serve a secondary purpose of allowing air pressure to equalize inside the tank when fuel is added or removed, while also preventing insects from flying or crawling into the vent piping and fouling the fuel in the tanks and pipes.

Safety

Flame arresters should be used only in the gas group and conditions they have been designed and tested for. Since the depth on an arrester is specified for certain conditions, changes in the temperature, pressure, or composition of the gases entering the arrester can cause the flame spatial velocity to increase, making the design of the arrester insufficient to stop the flame front ("propagation"). The deflagration may continue downstream of the arrester.

Flame arresters should be periodically inspected to make sure they are free of dirt, insects using it as a nest, or corrosion. The U.S. Chemical Safety and Hazard Investigation Board concluded that an uninspected and badly corroded flame arrester failed to prevent a 2006 explosion at a wastewater treatment plant in Daytona Beach, Florida.

Safety Valve

Safety valve is a valve that act as a protection of equipment from exploding or damaging and it is mainly installed in pressure vessels such as chemical plants, electric power boilers and gas storage tanks. Safety Valve is a type of valve that automatically actuates when the pressure of inlet side of the valve increases to a predetermined pressure, to open the valve disc and discharge the fluid (steam or gas) ; and when the pressure decreases to the prescribed value, to close the valve disc again. Safety valve is so-called a final safety device which controls the pressure and discharges certain amount of fluid by itself without any electric power support. Safety valve support not only the safety of energy industry but also the safety and security of our life.

Function and Design

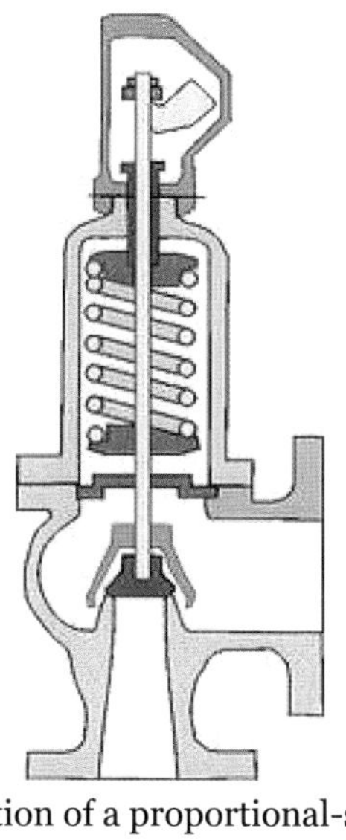

A cross-section of a proportional-safety valve

The earliest and simplest safety valve was used on a 1679 steam digester and utilized a weight to retain the steam pressure (this design is still commonly used on pressure cookers); however, these were easily tampered with or accidentally released. On the Stockton and Darlington Railway, the safety valve tended to go off when the engine hit a bump in the track. A valve less sensitive to sudden accelerations used a spring to contain the steam pressure, but these (based on a Salter spring balance) could still be screwed down to increase the pressure beyond design limits. This dangerous practice was sometimes used to marginally increase the performance of a steam engine. In 1856, John Ramsbottom invented a tamper-proof spring safety valve that became universal on railways. The Ramsbottom valve consisted of two plug-type valves connected to each other by a spring-laden pivoting arm, with one valve element on either side of the pivot. Any adjustment made to one of valves in an attempt to increase its operating pressure would cause the other valve to be lifted off its seat, regardless of how the adjustment was attempted. The pivot point on the arm was not symmetrically between the valves, so any tightening of the spring would cause one of the valves to lift. Only by removing and diassembling the entire valve assembly could its operating pressure be adjusted, making impromptu 'tying down' of the valve by locomotive crews in search of more power impossible. The pivoting arm was commonly extended into a handle shape lead back into the locomotive cab, allowing crews to 'rock' both valves off their seats to confirm they were set and operating correctly.

Safety valves also evolved to protect equipment such as pressure vessels (fired or not) and heat exchangers. The term safety valve should be limited to compressible fluid applications (gas, vapor, or steam).

The two general types of protection encountered in industry are thermal protection and flow protection.

For liquid-packed vessels, thermal relief valves are generally characterized by the relatively small size of the valve necessary to provide protection from excess pressure caused by thermal expansion. In this case a small valve is adequate because most liquids are nearly incompressible, and so a relatively small amount of fluid discharged through the relief valve will produce a substantial reduction in pressure.

Flow protection is characterized by safety valves that are considerably larger than those mounted for thermal protection. They are generally sized for use in situations where significant quantities of gas or high volumes of liquid must be quickly discharged in order to protect the integrity of the vessel or pipeline. This protection can alternatively be achieved by installing a high integrity pressure protection system (HIPPS).

Technical Terms

In the petroleum refining, petrochemical, chemical manufacturing, natural gas processing, power generation, food, drinks, cosmetics and pharmaceuticals industries, the term safety valve is associated with the terms pressure relief valve (PRV), pressure safety valve (PSV) and relief valve. The generic term is Pressure relief valve (PRV) or pressure safety valve (PSV) It should be noted that PRVs and PSVs are not the same thing, despite what many people think; the difference is that PSVs have a manual lever to open the valve in case of emergency.

- Relief valve (RV): an automatic system that is actuated by the static pressure in a liquid-filled vessel. It specifically opens proportionally with increasing pressure.
- Safety valve (SV): an automatic system that relieves the static pressure on a gas. It usually opens completely, accompanied by a popping sound.
- Safety relief valve (SRV): an automatic system that relieves by static pressure on both gas and liquid.

- Pilot-operated safety relief valve (POSRV): an automatic system that relieves on remote command from a pilot, to which the static pressure (from equipment to protect) is connected.
- Low pressure safety valve (LPSV): an automatic system that relieves static pressure on a gas. Used when the difference between the vessel pressure and the ambient atmospheric pressure is small.

Vacuum pressure safety valve (VPSV): an automatic system that relieves static pressure on a gas. Used when the pressure difference between the vessel pressure and the ambient pressure is small, negative and near to atmospheric pressure.

Low and vacuum pressure safety valve (LVPSV): an automatic system that relieves static pressure on a gas. Used when the pressure difference is small, negative or positive and near to atmospheric pressure.

RV, SV and SRV are spring-operated (even spring-loaded). LPSV and VPSV are spring-operated or weight-loaded.

Legal and Code Requirements in Industry

In most countries, industries are legally required to protect pressure vessels and other equipment by using relief valves. Also, in most countries, equipment design codes such as those provided by the ASME, API and other organizations like ISO must be complied with. These codes include design standards for relief valves and schedules for periodic inspection and testing after valves have been removed by the company engineer.

Today, the food, drinks, cosmetics, pharmaceuticals and fine chemicals industries call for hygienic safety valves, fully drainable and Cleanable-In-Place. Most are made of stainless steel; the hygienic norms are mainly 3A in the USA and EHEDG in Europe.

Development of The Safety Valve

Deadweight Lever Valves

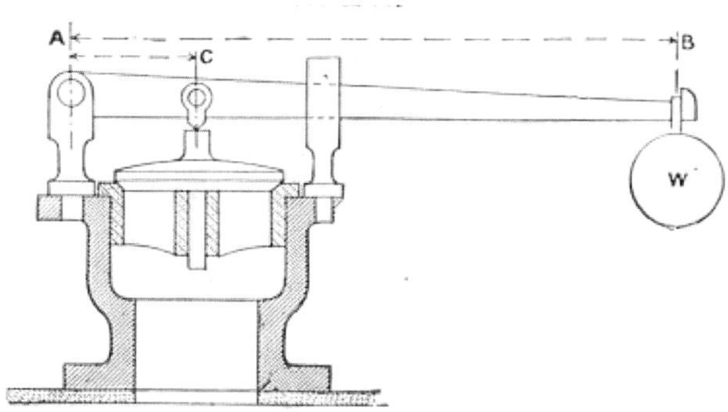

Lever arm safety valve

The first safety valve was invented by Denis Papin for his steam digester, an early pressure cooker rather than an engine. A weight acting through a lever held down a circular plug valve in the steam vessel. By using a "steelyard" lever a smaller weight was required, also the pressure could easily be regulated by sliding the same weight back and forth along the lever arm. Papin retained the same design for his 1707 steam pump. Early safety valves were regarded as one of the engineman's controls and required continuous attention, according to the load on the engine. In a famous early explosion at Greenwich in 1803, one of Trevithick's high-pressure stationary engines exploded when the boy trained to operate the engine left it to catch eels in the river, without first releasing the safety valve from its working load. By 1806, Trevithick was fitting pairs of safety valves, one external valve for the driver's adjustment and one sealed inside the boiler with a fixed weight. This was unadjustable and released at a higher pressure, intended as a guarantee of safety.

When used on locomotives, these valves would rattle and leak, releasing near-continuous puffs of waste steam.

Direct-Acting Deadweight Valves

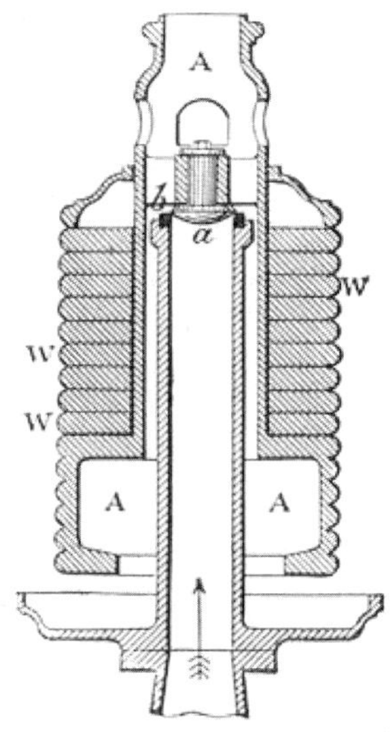

Deadweight safety valve.

Although the lever safety valve was convenient, it was too sensitive to the motion of a steam locomotive. Early steam locomotives therefore used a simpler arrangement of weights stacked directly upon the valve. This required a smaller valve area, so as to keep the weight manageable, which sometimes proved inadequate to vent the pressure of an unattended boiler, leading to explosions. An even greater hazard was the ease with which such a valve could be tied down, so as to increase the pressure and thus power of the engine, at further risk of explosion.

Although deadweight safety valves had a short lifetime on steam locomotives, they remained in use on stationary boilers for as long as steam power remained.

Direct Spring Valves

Locomotive Planet, with a brass-cased direct spring valve

Weighted valves were sensitive to bouncing from the rough riding of early locomotives. One solution was to use a lightweight spring rather than a weight. This was the invention of Timothy Hackworth on his Royal George of 1828. Owing to the limited metallurgy of the period, Hackworth's first spring valves used an accordion-like stack of multiple leaf springs.

These direct-acting spring valves could be adjusted by tightening the nuts retaining the spring. To avoid tampering, they were often shrouded in tall brass casings which also vented the steam away from the locomotive crew.

Salter Spring Balance Valves

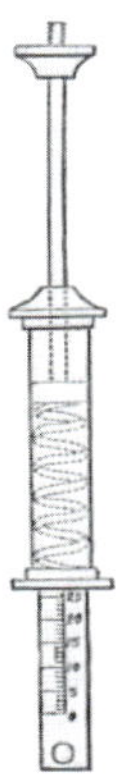

Salter Spring Balance

The Salter coil spring spring balance for weighing, was first made in Britain by around 1770. This used the newly developed spring steels to make a powerful but compact spring in one piece. Once again by using the lever mechanism, such a spring balance could be applied to the considerable force of a boiler safety valve.

The spring balance valve also acted as a pressure gauge. This was useful as previous pressure gauges were unwieldy mercury manometers and the Bourdon gauge had yet to be invented.

Lockable Valves

Phoenix with two sets of Salter spring balance valves

The risk of firemen tying down the safety valve remained. This was encouraged by them being fitted with easily adjustable wing nuts, the practice of adjusting the boiler's working pressure via the safety valve being an accepted behaviour well into the 1850s. It was later common with Salter valves for them to be fitted in pairs, one adjustable and often calibrated for use as a gauge, the other sealed inside a locked cover to prevent tampering.

Paired Spring Balance Valves

Midland Spinner, showing paired spring-balance safety valves behind the dome

Paired spring-balance safety valves of a ČSD Class 252.0, with hand adjustment wheels

Paired valves were often adjusted to slightly different pressures too, a small valve as a control measure and the lockable valve made larger and permanenently set to a higher pressure, as a safeguard. Some designs, such as one by Sinclair for the Eastern Counties Railway in 1859, had the valve spring with pressure scale behind the dome, facing the cab, and the locked valve ahead of the dome, out of reach of interference.

Ramsbottom safety valves

Ramsbottom valves on a model traction engine

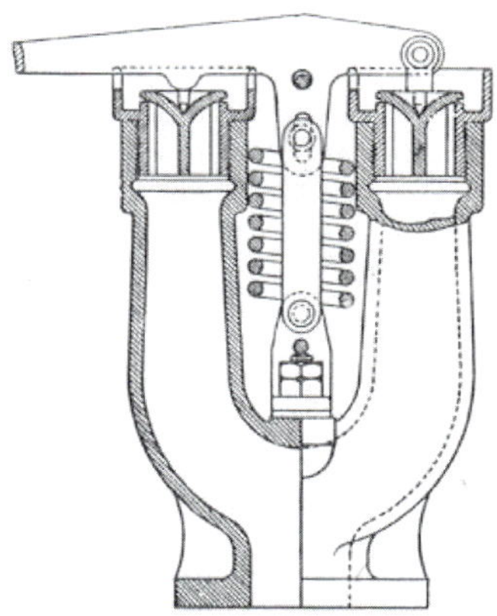

U-shaped Ramsbottom safety valve

In 1855, John Ramsbottom, later locomotive superintendent of the LNWR, described a new form of safety valve intended to improve reliability and especially to be tamper-resistant. A pair of plug valves were used, held down by a common spring-loaded lever between them with a single central spring. This lever was characteristically extended rearwards, often reaching into the cab on early locomotives. Rather than discouraging the use of the spring lever by the fireman, Ramsbottom's valve encouraged this. Rocking the lever freed up the valves alternately and checked that neither was sticking in its seat. Even if the fireman held the lever down and increased the force on the rear valve, there was a corresponding reduction of force on the forward valve.

Various forms of Ramsbottom valve were produced. Some were separate fittings to the boiler, through separate penetrations. Others were contained in a U-shaped housing fastened to a single opening in the boiler shell. As boiler diameter increased, some forms were even set inside the boiler shell, with the springs housed in a recess inside and only the valves and balance lever protruding outside. These had obvious drawbacks for easy maintenance.

A drawback to the Ramsbottom type was its complexity. Poor maintenance or mis-assembly of the linkage between the spring and the valves could lead to a valve that no longer opened correctly under pressure. The valves could be held against their seats and fail to open or, even worse, to allow the valve to open but insufficiently to vent steam at an adequate rate and so not being an obvious and noticeable fault. Mis-assembly of just this nature led to a fatal boiler explosion in 1909 at Cardiff on the Rhymney Railway, even though the boiler was almost new, at only eight months old.

Naylor valves were introduced around 1866. A bellcrank arrangement reduced the strain (percentage extension) of the spring, thus maintaining a more constant force. They were used by the L&Y & NER.

"Pop" Valves

Ross pop valve, from Tornado

All of the preceding safety valve designs opened gradually and had a tendency to leak a "feather" of steam as they approached "blowing-off", even though this was below the pressure. When they opened they also did so partially at first and didn't vent steam quickly until the boiler was well over pressure.

The quick-opening "pop" valve was a solution to this. Their construction was simple: the existing circular plug valve was changed to an inverted "top hat" shape, with an enlarged upper diameter. They fitted into a stepped seat of two matching diameters. When closed, the steam pressure acted only on the crown of the top hat, and was balanced by the spring force. Once the valve opened a little, steam could pass the lower seat and began to act on the larger brim. This greater area overwhelmed the spring force and the valve flew completely open with a "pop". Escaping steam on this larger diameter also held the valve open until pressure had dropped below that at which it originally opened, providing hysteresis.

Cover of The Pop Valve

These valves coincided with a change in firing behaviour. Rather than demonstrating their virility by always showing a feather at the valve, firemen now tried to avoid noisy blowing off, especially around stations or under the large roof of a major station. This was mostly at the behest of stationmasters, but firemen also realised that any blowing off through a pop valve wasted several pounds of boiler pressure; estimated at 20 psi lost and 16 lbs or more of shovelled coal.

GWR safety valve cover

Pop valves derived from Adams's patent design of 1873, with an extended lip. R L Ross's valves were patented in 1902 & 1904. They were more popular in America at first, but widespread from 1920s.

Although showy polished brass covers over safety valves had been a feature of steam locomotives since Stephenson's day, the only railway to maintain this tradition into the era of pop valves was the GWR, with their distinctive tapered brass safety valve bonnets and copper-capped chimneys.

Marine and Cockburn High-Lift Safety Valves

Developments in high-pressure water-tube boilers for marine use placed more demands on safety valves. Valves of greater capacity were required, to vent safely the high steam-generating capacity of these large boilers. As the force on their valves increased, the issue of the spring's increasing stiffness as its load increased (like the Naylor valve) became more critical. The need to reduced valve feathering became even more important with high-pressure boilers, as this represented both a loss of distilled feedwater and also a scouring of the valve seats, leading to wear.

High-lift safety valves are direct-loaded spring types, although the spring does not bear directly on the valve, but on a guide-rod valve stem. The valve is beneath the base of the stem, the spring rests on a flange some height above this. The increased space between the valve itself and the spring seat allows the valve to lift higher, further clear of the seat. This gives a steam flow through the valve equivalent to a valve one and a half or twice as large (depending on detail design).

The Cockburn Improved High Lift design has similar features to the Ross pop type. The exhaust steam is partially trapped on its way out and acts on the base of the spring seat, increasing the lift force on the valve and holding the valve further open.

To optimise the flow through a given diameter of valve, the full-bore design is used. This has a servo action, where steam through a narrow control passage is allowed though if it passes a small control valve. This steam is then not exhausted, but is passed to a piston that is used to open the main valve.

There are safety valves known as PSV's and can be connected to pressure gauges (usually with a 1/2" BSP fitting). These allow a resistance of pressure to be applied to limit the pressure forced on the gauge tube, resulting in prevention of over pressurisation. the matter that has been injected into the gauge, if over pressurised, will be diverted through a pipe in the safety valve, and shall be driven away from the gauge.

Types

Steam locomotive No. 46229, Duchess of Hamilton lifts her boiler safety valve after hauling the Welsh Marches Pullman charter.

There is a wide range of safety valves having many different applications and performance criteria in different areas. In addition, national standards are set for many kinds of safety valves.

United States

- ASME (American Society of Mechanical Engineers) Boiler & Pressure Vessel Code, Section I
- ASME (American Society of Mechanical Engineers) Boiler & Pressure Vessel Code, Section VIII, Division 1
- API (American Petroleum Institute) Recommended Practice 520 and API Standard 526, API Standard 2000 (low pressure - Storage tank)

European Union

European standard steam boiler safety valve:

- ISO 4126 (harmonized with European Union directives).

- EN 764-7 (former CEN standard, harmonized with European Union directives, replaced with EN ISO 4126-1).
- AD Merkblatt (German).
- PED 97/23/CE (Pressure Equipment Directive - European Union).

Water Heaters

Temperature and Pressure safety valve on a water heater.

Safety valves are required on water heaters, where they prevent disaster in certain configurations in the event that a thermostat should fail. There are still occasional, spectacular failures of older water heaters that lack this equipment. Houses can be leveled by the force of the blast.

Pressure Cookers

Pressure cookers are cooking pots with a pressure-proof lid. Cooking at pressure allows the temperature to rise above the normal boiling point of water (100 degrees Celsius at sea level), which speeds up the cooking and makes it more thorough.

Pressure cookers usually have two safety valves to prevent explosions. On older designs, one is a nozzle upon which a weight sits. The other is a sealed rubber grommet which is ejected in a controlled explosion if the first valve gets blocked. On newer generation pressure cookers, if the steam vent gets blocked, a safety spring will eject excess pressure and if that fails, the gasket will expand and release excess pressure downwards between the lid and the pan. Also, newer generation pressure cookers have a safety interlock which locks the lid when internal pressure exceeds atmospheric pressure, to prevent accidents from a sudden release of very hot steam, food and liquid, which would happen if the lid were to be removed when the pan is still slightly pressurised inside (however, the lid will be very hard or impossible to open when the pan is still pressurised).

Explosion-proof Enclosures

Explosion-proof enclosures are cabinets that contain varied electrical equipment like transformers, switches, controls, plugs and knobs that prevent electrical fires, explosions, shocks and other electrical hazards. These enclosures are designed to be spark and shock resistant and also have a high tolerance to high temperatures. These are an ideal safety solution in locations that house crucial electrical equipment and can reduce risks to life, health and property.

Types of Explosion-Proof Enclosures

Explosion-proof enclosures are classified into different types depending on the location and the level of protection they offer. These codes are based on National Electrical Manufacturers Association (NEMA) standards which indicate the level of protection against electrical hazards like fire and explosion. Below is a list of location-specific enclosures and their applications:

Non-hazardous Locations

Type 1

Designed for indoor use, type 1 enclosures offer protection against hazardous parts and protect the electronic equipment from dirt, dust, and debris.

Type 2

These enclosures are constructed for indoor applications and provide protection to the employees from electric hazards. It keeps the electrical equipment protected from the harmful effects of foreign objects.

Type 3

Designed for both indoor and outdoor use, type 3 enclosures keep the personnel protected from hazardous parts and the equipment safe from foreign objects and ingress of sleet, snow, dust, dirt, and rain.

Type 3R

In addition to the features and benefits that accompany Type 1 and Type 2, this make offers a much higher resistance to rain, sleet, snow, dust, and extreme weather conditions.

Type 3S

Offers protection against hazardous parts and guards the equipment from getting damaged by solid foreign objects. Type 3S enclosures can be used both indoors and outdoors to keep electrical equipment from deteriorating due to the ingress of rain, snow and sleet and functional flaws even when it is fully covered with ice.

Type 3X

Over and above the protection offered by Type 3R, these enclosures offer a much higher degree of protection against the external environment. The designation X stands for resistance to corrosion.

Type 3RX

Constructed for both indoor and outdoor use, this specific type offers additional protection against snow, sleet and extreme weather conditions apart from being corrosion resistant.

Type 3SX

It has all the features and benefits that accompany Type 3S. Designed for rugged use, this make features an efficient external mechanism that is corrosion resistant and facilitates operation even when the unit is covered with snow.

Types 4, 4X

These enclosures are ideal for both indoor and outdoor use and designed for protection against hazardous parts. These units keep the inside equipment guarded against harsh environmental conditions and ingress of external factors like rain, snow, dirt, dust and sleet. The designation X signifies resistance to corrosion which implies that the enclosure remains undamaged even with the formation of ice.

Type 5

Type 5 enclosures feature a mechanism that is similar to type 2. It comes with an additional gasket that keeps the enclosure guarded against fibers, lint, dust, dirt, and flyings. It is ideal for indoor use.

Types 6, 6P

Over and above the safety offered by type 4 units, these have a much higher resistance to water even when there are chances of temporary immersion. It prevents the entry of water and keeps hazardous consequences at bay in case of prolonged immersion.

Types 12, 12K

Designed specifically for indoor use, these enclosures provide a very high level of protection against dust, dirt, debris, lint, drips, water splashes, fibers, and flyings. Type 12 units are not resistant to rusting and corrosion but 12k units are. 12k enclosures are available in multiple variants including heavy-duty, wall mount, floor mount, modular, consoles, pushbutton and more.

Type 13

These all-purpose units are specifically constructed for indoor use and provide complete protection against ingress of circulating dust, blowing dirt, debris, lint, fibers, and flyings. It keeps the equipment protected against seepage of non-corrosive coolants, splashing, oil, and dripping.

Explosion-proof enclosures rated for hazardous locations

Type 7 Enclosures

Type 7 enclosures are designed for indoor use in Division 1, Class 1, and Groups A, B, C, and D as defined by the National Fire Protection Association.

Type 8 Enclosures

Type 8 enclosures are designed to meet the needs of indoor as well as outdoor applications in hazardous locations. These units are compatible for use in Division 1, Class 1 and Groups A, B, C, and D.

Type 9 Enclosures

Compatible with indoor and outdoor installations in hazardous locations, these units are an ideal solution for Division 1, Class 1, and Groups E, F and G.

Type 10 Enclosures

Type 10 enclosures are designed to meet the safety specifications of Mines and Health Administration as stated under part 18 of CFR 30. Environments that contain combustible substances need multi-rated explosion-proof enclosures. Certain environments may also need high pressurizing and purging, provided there is a dependable system that complies with the safety requirements of National Fire Protection Association (NFPA).

Applications of Explosion-Proof Enclosures

Explosion-proof enclosures are available in different shapes, sizes, and specifications for different applications. These spark resistant units house electrical equipment and ensure complete safety of surroundings. Typical applications include:

- Petroleum refineries
- Industries using flammable liquids for operations

- Petrochemical industries
- Dry cleaning plants
- Fuel servicing areas
- Gas plants
- Feed mills
- Plants running metallurgical processes
- Chemical plants
- Producers of plastics and fireworks
- Producers of medicines, sugar plants, and candies
- Carbon-handling units
- Textile mills
- Flax processing plants

Choosing Enclosures for Industrial Applications

With more and more companies going green, electronic products are now being designed for use in turbines, solar power systems and alternative energy systems that run with the aid of power and employ smart grids. Faulty installations can lead to fatal injuries and accidents. With the increasing need for electronics in every industry, it is important to consider these five points when selecting enclosures for specific applications:

- Location

Where the product will be installed determines the level of protection needed. Wind turbines are installed outdoors due to which they need to meet NEMA specifications for which 4X enclosures need to be installed. It is important to determine which enclosure has the required strength and durability to protect the equipment from dripping water, powerful sprays, dirt, dust, snow, sleet and debris depending on the location of installation.

- Capabilities

The capacity of your chosen enclosure plays a vital role in maintaining safety and speeding installation. The availability of brackets and attachments can be a huge help. It is important to have a mounting panel when installing a variety of electric components. Standard enclosures are readily available and cost much less as compared to customized units.

- NEMA/RFI Ratings

Applications that involve monitoring and consolidation of excess power from different sources require added protection. A die-cast enclosure made from aluminum with a double layering of gaskets and an outer gasket for sealing the moisture makes an ideal solution for such applications.

- Dimensions

Apart from environmental safety, you should also consider the future needs of your industry. The scope and extent of expansion play a decisive role in determining the dimensions of the enclosure. A small enclosure may prove to be an excellent space saver but can invite significant expenses in the long run. Enclosure modifications for cable access also need to be taken into consideration in enclosure selection.

- Design

The appearance of the enclosure becomes particularly important when it is being installed in a home or an office. The material of the enclosure also needs to be selected depending on the end-use. Installing an industrial unit for controlling a domestic solar system is not a sound choice. A plastic unit with a sleek design makes a perfect solution for residential installations.

Insulation Monitoring Device

An insulation monitoring device monitors the ungrounded system between an active phase conductor and earth. It is intended to give an alert (light and sound) or disconnect the power supply when the resistance between the two conductors drops below a set value, usually 50 kΩ (sample of IEC standard for medical applications). The main advantage is that the ungrounded or floating system allows a continuous operation of important consumers such as medical, chemical, military, etc.

Some manufacturers of monitors for these systems are capable of handling VFDs (Variable Frequency (Speed) Drives). Most, however are not due to issues with the DC-portions of the VFDs.

Most monitors work by injecting low level dc on the line and detecting. Some manufacturers use a patented AMP-monitoring principle (Adapted Measuring Pulse)

Chemical Protective Clothing

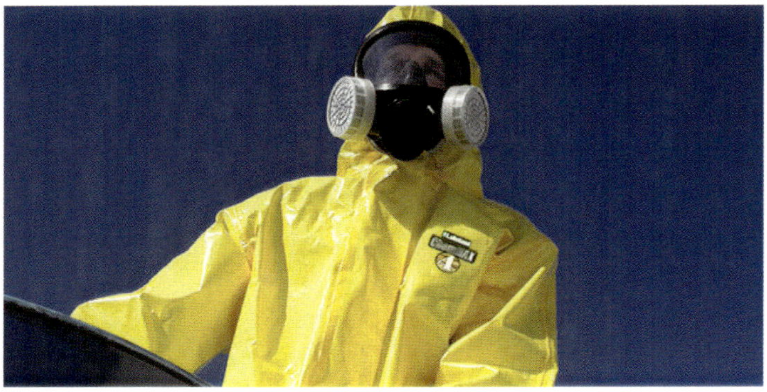

Chemical protective clothing is workplace protective clothing used to protect the worker from exposure to chemicals. It should not be viewed as a replacement for proper handling and safety engineering methods.

References

- Arpentinier, Philippe; Cavani, Fabrizio; Trifiro, Ferrucio (2001). The Technology of Catalytic Oxidations. 2. Paris: Editions TECHNIP. p. 748. ISBN 2-7108-0777-7. Retrieved November 19, 2007.
- Blowout-preventer-bop-6083: petropedia.com, Retrieved 16 April 2018
- "Deepwater Horizon Joint Investigation Team official Web site". U.S. Coast Guard and Minerals Management Service. Retrieved 2010-05-26
- Flame-arrester, equipment: wermac.org, Retrieved 25 March 2018
- McManus, Neil (1998). Safety and Health in Confined Spaces. CRC Press. p. 147. ISBN 1-56670-326-3. Retrieved November 19, 2007
- Chemical-protective-clothing-4890: safeopedia.com, Retrieved 29 July 2018
- Clark, Andrew (2010-06-18). "BP oil disaster puts spotlight on small Texan firm". The Guardian. Retrieved 19 June 2010

Permissions

All chapters in this book are published with permission under the Creative Commons Attribution Share Alike License or equivalent. Every chapter published in this book has been scrutinized by our experts. Their significance has been extensively debated. The topics covered herein carry significant information for a comprehensive understanding. They may even be implemented as practical applications or may be referred to as a beginning point for further studies.

We would like to thank the editorial team for lending their expertise to make the book truly unique. They have played a crucial role in the development of this book. Without their invaluable contributions this book wouldn't have been possible. They have made vital efforts to compile up to date information on the varied aspects of this subject to make this book a valuable addition to the collection of many professionals and students.

This book was conceptualized with the vision of imparting up-to-date and integrated information in this field. To ensure the same, a matchless editorial board was set up. Every individual on the board went through rigorous rounds of assessment to prove their worth. After which they invested a large part of their time researching and compiling the most relevant data for our readers.

The editorial board has been involved in producing this book since its inception. They have spent rigorous hours researching and exploring the diverse topics which have resulted in the successful publishing of this book. They have passed on their knowledge of decades through this book. To expedite this challenging task, the publisher supported the team at every step. A small team of assistant editors was also appointed to further simplify the editing procedure and attain best results for the readers.

Apart from the editorial board, the designing team has also invested a significant amount of their time in understanding the subject and creating the most relevant covers. They scrutinized every image to scout for the most suitable representation of the subject and create an appropriate cover for the book.

The publishing team has been an ardent support to the editorial, designing and production team. Their endless efforts to recruit the best for this project, has resulted in the accomplishment of this book. They are a veteran in the field of academics and their pool of knowledge is as vast as their experience in printing. Their expertise and guidance has proved useful at every step. Their uncompromising quality standards have made this book an exceptional effort. Their encouragement from time to time has been an inspiration for everyone.

The publisher and the editorial board hope that this book will prove to be a valuable piece of knowledge for students, practitioners and scholars across the globe.

Index